Visibilities and Invisibilities in Smart Cities:

Emerging Research and Opportunities

H. Patricia McKenna
AmbientEase, Canada

A volume in the Advances in Civil and Industrial Engineering (ACIE) Book Series

IGI Global
PUBLISHER of TIMELY KNOWLEDGE

Published in the United States of America by
IGI Global
Engineering Science Reference (an imprint of IGI Global)
701 E. Chocolate Avenue
Hershey PA, USA 17033
Tel: 717-533-8845
Fax: 717-533-8661
E-mail: cust@igi-global.com
Web site: http://www.igi-global.com

Library of Congress Cataloging-in-Publication Data

Names: McKenna, H. Patricia, 1954- author.
Title: Visibilities and invisibilities in smart cities : emerging research and opportunities / by H. Patricia McKenna.
Description: Hershey, PA : Engineering Science Reference, [2021] | Includes bibliographical references and index. | Summary: "This authored book is concerned with understandings of, and perspectives on, the combination of visible and invisible elements that contribute to the making of smart cities"-- Provided by publisher.
Identifiers: LCCN 2020042264 (print) | LCCN 2020042265 (ebook) | ISBN 9781799838500 (h/c) | ISBN 9781799872498 (s/c) | ISBN 9781799838517 (eISBN)
Subjects: LCSH: Smart cities. | Streetscapes (Urban design) | City and town life--Research.
Classification: LCC TD159.4 .M385 2021 (print) | LCC TD159.4 (ebook) | DDC 307.1/160285--dc23
LC record available at https://lccn.loc.gov/2020042264
LC ebook record available at https://lccn.loc.gov/2020042265

This book is published in the IGI Global book series Advances in Civil and Industrial Engineering (ACIE) (ISSN: 2326-6139; eISSN: 2326-6155)

British Cataloguing in Publication Data
A Cataloguing in Publication record for this book is available from the British Library.

For electronic access to this publication, please contact: eresources@igi-global.com.

Advances in Civil and Industrial Engineering (ACIE) Book Series

ISSN:2326-6139
EISSN:2326-6155

Editor-in-Chief: Ioan Constantin Dima, University Valahia of Târgoviște, Romania

MISSION

Private and public sector infrastructures begin to age, or require change in the face of developing technologies, the fields of civil and industrial engineering have become increasingly important as a method to mitigate and manage these changes. As governments and the public at large begin to grapple with climate change and growing populations, civil engineering has become more interdisciplinary and the need for publications that discuss the rapid changes and advancements in the field have become more in-demand. Additionally, private corporations and companies are facing similar changes and challenges, with the pressure for new and innovative methods being placed on those involved in industrial engineering.

The **Advances in Civil and Industrial Engineering (ACIE) Book Series** aims to present research and methodology that will provide solutions and discussions to meet such needs. The latest methodologies, applications, tools, and analysis will be published through the books included in **ACIE** in order to keep the available research in civil and industrial engineering as current and timely as possible.

COVERAGE

- Earthquake engineering
- Ergonomics
- Transportation Engineering
- Hydraulic Engineering
- Productivity
- Structural Engineering
- Optimization Techniques
- Urban Engineering
- Quality Engineering
- Operations Research

IGI Global is currently accepting manuscripts for publication within this series. To submit a proposal for a volume in this series, please contact our Acquisition Editors at Acquisitions@igi-global.com or visit: http://www.igi-global.com/publish/.

The Advances in Civil and Industrial Engineering (ACIE) Book Series (ISSN 2326-6139) is published by IGI Global, 701 E. Chocolate Avenue, Hershey, PA 17033-1240, USA, www.igi-global.com. This series is composed of titles available for purchase individually; each title is edited to be contextually exclusive from any other title within the series. For pricing and ordering information please visit http://www.igi-global.com/book-series/advances-civil-industrial-engineering/73673. Postmaster: Send all address changes to above address.

Titles in this Series

For a list of additional titles in this series, please visit:
http://www.igi-global.com/book-series/advances-civil-industrial-engineering/73673

Transforming Urban Nightlife and the Development of Smart Public Spaces
Hisham Abusaada (Housing and Building National Research Center (HBRC), Egypt) Abeer Elshater (Ain Shams University, Egypt) and Dennis Rodwell (Independent Reseacher, UK)
Information Science Reference • ©2021 • 290pp • H/C (ISBN: 9781799870043) • US $195.00

Handbook of Research on Driving Transformational Change in the Digital Built Environment
Jason Underwood (University of Salford, UK) and Mark Shelbourn (Birmingham City University, UK)
Engineering Science Reference • ©2021 • 400pp • H/C (ISBN: 9781799866008) • US $295.00

Structural Dynamics and Static Nonlinear Analysis From Theory to Application
Mourad Belgasmia (Setif 1 University, Algeria)
Engineering Science Reference • ©2021 • 347pp • H/C (ISBN: 9781799843993) • US $195.00

Handbook of Research on Developing Smart Cities Based on Digital Twins
Matteo Del Giudice (Politecnico di Torino, Italy) and Anna Osello (Politecnico di Torino, Italy)
Engineering Science Reference • ©2021 • 674pp • H/C (ISBN: 9781799870913) • US $295.00

Data-Driven Optimization of Manufacturing Processes
Kanak Kalita (Vel Tech Rangarajan Dr. Sagunthala R&D Institute of Science and Technology, India) Ranjan Kumar Ghadai (Sikkim Manipal Institute of Technology, Sikkim Manipal University, Majhitar, India) and Xiao-Zhi Gao (University of Eastern Finland, Finland)
Engineering Science Reference • ©2021 • 298pp • H/C (ISBN: 9781799872061) • US $225.00

For an entire list of titles in this series, please visit:
http://www.igi-global.com/book-series/advances-civil-industrial-engineering/73673

IGI Global
PUBLISHER of TIMELY KNOWLEDGE

701 East Chocolate Avenue, Hershey, PA 17033, USA
Tel: 717-533-8845 x100 • Fax: 717-533-8661
E-Mail: cust@igi-global.com • www.igi-global.com

Table of Contents

Section 4
Ambient Invisibilities Going Forward

Preface

Visibilities and Invisibilities in Smart Cities: Emerging Research and Opportunities focuses on people becoming more in tune with their awareness capabilities through their experiences and assessments of everyday interactions and activities in the context of more technologically aware urban environments and regions. Specifically, this work is concerned with understandings of, and perspectives on, the combination of visible and invisible elements that contribute to the making of smart cities. In other words, how do people make sense at the urban level and in extra-urban spaces, of the combined complexities of invisibilities and visibilities in their environments, interactions, and infrastructures enabled through their own enhanced awareness together with technologies characterized by increasing awareness capabilities that are often embedded, pervasive, and ambient? In response, this book probes the visible and invisible dimensions of emerging understandings of smart cities and regions in the context of more aware people interacting with each other and through more aware and pervasive technologies. Where earlier works identified variables associated with visibilities and invisibilities in smart cities, this work explores the relationships between variables using a case study approach combined with an explanatory correlational design. This work is based on an urban research study conducted from mid-2015 into early 2021 spanning multiple countries (e.g., Canada, the United States, several European countries, and Israel) across three continents (e.g., North America, Europe, and the Middle East). As such, this book makes a contribution to the research literature for smart cities; the evolving and enriching of spaces for urban theory; and the evolving and enriching of methodologies, approaches, and inquiries in support of research and practice pertaining to smarter and more responsive cities. This work also advances the formulation of a conceptual framework for visibilities and invisibilities in smart cities as a resource for evolving spaces for theory, research, and practice. As such, this work is intended to provide explorations of, and insights into, rapidly emerging transformations occurring in urban areas and regions and in society more generally. Such transformations are associated with the embedded, pervasive, continuous, and dynamic nature of information and communication technologies (ICTs), computational intelligence (CI), computing networks (CN), and other emerging technologies that this work refers to as *awareness-enabled and enabling technologies* exhibiting awareness capabilities or augmentations, pertaining to location, context, and so on.

This work came about from an interest in expanding upon theory and methods for the notion of visibilities and invisibilities in Smart Cities. Crucial to this body of work is the importance of people as critical to the urban fabric of cities and more particularly, to evolving understandings of smart cities and regions as we enter the third decade of the 21st century. As such, the purpose of this manuscript is to offer a lens through which to view and explore perspectives on smart cities. For example, the potential for transparency and translucence as dimensions of visibilities and invisibilities emerges, enabling insights into the complexities and seemingly intractable challenges. And yet, opportunities are offered in the making of smarter and more responsive cities, based on evolving 21st century needs, innovations, and transformations. In turn, this work explores physical and digital elements underpinning urban visibilities and invisibilities, through the probing of notions of transparency and translucence. This exploratory and explanatory work responds with an interdisciplinary approach to the need for extending methodologies and spaces for theory in understanding the importance of elements in urban and extra-urban environments and regions informing smart, responsive, and future cities. The research literature for urban theory and urban methodologies is explored in this work in the context of smarter and more aware cities, as in, ambient, enabling the identification of opportunities for research and practice associated with urban visibilities and invisibilities.

This book will have interdisciplinary appeal across multiple domains including, but not limited to, architecture, environmental design, human-computer interaction, human geography, information technology, sociology, and affective computing, to name a few. This work is intended for students, educators, business innovators, community innovators, designers, artists, urban innovators, city officials and staff, policymakers, and anyone concerned with imagining and designing more livable cities and urban environments and regions. Designed as a guide for evolving pathways for urban studies and practices, this book will be of interest to educators as a course textbook; to researchers for further testing and validation of variables, correlations, and approaches presented throughout the book; and to urban practitioners in finding ways of involving people more meaningfully in smarter urban initiatives, influencing policymaking, and transforming the city for improved livability.

This book consists of eight chapters arranged within four sections and the organization is designed to provide: an introduction to the concepts of the visible and the invisible generally and in relation to aspects of smart cities; frameworks for understanding the interplay and interweaving of these two concepts as InVisibilities; associated and evolving theory and methods for study and practices around InVisibilities; along with the extending of current thinking on emerging and evolving research on invisibilities and visibilities and the ambient as opportunities in smart city environments and regions. A brief description of each section and chapter in the book is provided below.

The purpose of Section 1 is to provide a review of the research literature for the visibilities and invisibilities concepts in the context of smart cities in Chapters 1 and 2. Chapter 1 focuses on identifying and articulating what contributes to the visible and invisible dimensions of urban spaces and regions in smart cities through an introduction to the concepts of visibilities and invisibilities and what constitutes the visible and invisible in the research literature generally and in relation to smart cities more particularly. For example, it could be said that the visible interaction of people talking to each other on a city street is rendered invisible as people use digital communications, engage with social media, and the like. A conceptual framework is advanced for visibilities and invisibilities in smart cities and regions and operationalized for use in the chapter. Information and communication technologies (ICTs) are used as a proxy for invisible infrastructures to explore the nature of the relationship with visible infrastructures. A key takeaway from this chapter is the presence of a promising relationship between ICTs as a proxy for invisible infrastructures when correlated with visible infrastructures in smart cities. The correlation is then enriched with qualitative data highlighting aspects of technology infrastructures and human infrastructures and examples of invisible infrastructures (e.g., from data visualizations to sensing air quality) and visible infrastructures (e.g., urban displays).

Chapter 2 provides an introduction to the notion of making the invisible visible in smart cities using techniques such as imagination, simulation, discourse, and art, to name a few. In doing so, this chapter identifies and articulates what contributes to making some of the invisible dimensions of urban spaces and regions visible. For example, through data visualizations, the often-invisible nature of interactions between people communicating with each other in urban environments is rendered visible, shedding light on data potentials in addition to when, where, and to what extent communications occur. Approaches to making the invisible visible discussed in this chapter point to an array of methodologies amenable to studies and practices in the smart cities domain while advancing a conceptual framework for making the invisible visible in smart cities. Practices for making the invisible visible are explored through looking at the nature of the relationship between heightening urban sensibilities and urbanizing (e.g., adapting for urban uses). The exploration is extended to assessments of emotion/affect in terms of people and their experience of tension in urban environments such anger (representing InVisibilities) in relation to awareness (representing the ambient). A key takeaway from this chapter is the presence of a "useful" correlation between heightening urban sensibilities as a proxy for invisibilities and urbanizing as in, adapting for urban uses, as a proxy for visibilities in smart cities. The correlation is then enriched with qualitative data highlighting the augmenting of urban services with the use of drones and the like.

Section 2 explores the dimensions of transparency, translucence, and the ambient in extending understandings of visibilities and invisibilities in smart cities in Chapters 3 and 4. Chapter 3 addresses the notion of what constitutes transparency and more particularly, in relation to smart cities while identifying emerging research and opportunities through an exploration of transparency in term of data, design, and openness. Also explored is the nature of the ambient in the context of smart cities. A framework for transparency and the ambient in smart cities is advanced and operationalized for use in the chapter. The relationship between transparency (using openness as a proxy) and the ambient (using attuning to urban spaces as a proxy) is explored in smart cities. A key takeaway from this chapter is the presence of a "slight relationship" between openness as a proxy for transparency when correlated with attuning to urban spaces as a proxy for the ambient in smart cities. The correlation is then enriched with qualitative data highlighting urban initiatives pertaining to data, design, and experiences that take people and their needs into consideration.

Chapter 4 explores the notion of translucence generally, and in relation to smart cities through a review of the literature for social translucence, interface translucence, building materials, and learning. The nature of the ambient in the context of smart cities is also explored. Translucence and the ambient are then explored in relation to healthcare, ambient experience, and ambient computing. A framework for translucence and the ambient in smart cities is formulated and operationalized for use in the chapter. Based on the framework, the relationship between translucence (using visible infrastructures as a proxy) and the ambient (using awareness as a proxy) is explored in smart cities and why such a relationship matters for research and practice. A key takeaway from this chapter is the presence of a "slight relationship" between translucence using visible infrastructures as a proxy when correlated with the ambient using awareness as a proxy in smart cities. The correlation is then enriched with qualitative data highlighting concerns with accountability, trust, constraints, and the social, taking the needs of people into consideration.

With this background, Section 3 addresses theory and methods in urban environments and regions in relation to visibilities and invisibilities in Chapters 5 and 6, respectively. Chapter 5 explores theory pertaining to atmospheres, cities, change, complexity, and constraints in support of perspectives on invisibilities and visibilities, and their interwoven-ness as InVisibilities, in smart cities and regions. A framework for InVisibilities and the ambient is advanced in support of evolving spaces for theory in smart cities and operationalized for use in the chapter. The notion of ambient InVisibilities in smart cities is explored using information and communication technologies (ICTs) as a proxy for InVisibilities and awareness as a proxy for the ambient. A key takeaway from this chapter is the presence of a "slight" inverse relationship between ICTs as a proxy for InVisibilities when correlated with awareness as a proxy for the ambient in smart cities. The correlation is then enriched with qualitative data highlighting atmospheres in relation to the pulse, vibrancy, and sense of the city.

Chapter 6 expands upon methodological spaces and approaches for the exploring of visibilities and invisibilities in smart cities in the context of the emerging challenges and opportunities for both practice and research. Methods explored include appreciative inquiry, datawalking, design thinking, simulation, imagination, and space syntax, to name a few. As such, this chapter places an emphasis on methods and approaches in support of working with InVisibilities in smart cities. For example, this chapter argues for extending the emerging *self-relevant research* approach by Amabile and Hall (2019) to the study of smart cities. A framework for methodologies accommodating spaces for visibilities and invisibilities in smart cities is advanced. The framework is operationalized for use in the chapter where placemaking as a proxy for InVisibilities is explored in relation to attuning to urban spaces as a proxy for the ambient. A key takeaway from this chapter is the presence of "limited prediction" in the relationship that emerged between placemaking as a proxy for InVisibilities when correlated with attuning to urban spaces as a proxy for the ambient in smart cities. The correlation is then enriched with qualitative data highlighting the importance of approaches to fostering education, learning, and engagement in urban environments.

Section 4 offers a look at ambient InVisibilities going forward in urban environments and regions in Chapters 7 and 8. Chapter 7 provides a look to the future with a discussion of ambient approaches and inquiries with considerations of well-being and livability as a way of involving people more directly in understandings and assessments of InVisibilities in smart cities and regions. A review of the research literature for emergent complexities is presented such as that of the global COVID-19 pandemic. Literature is then considered in relation to the framing of ambient approaches and inquiries for InVisibilities in smart cities. The conceptual framework for ambient InVisibilities is extended in Chapter 7 to form a conceptual framework for augmented awareness for ambient InVisibilities in smart cities and regions. As such, this chapter explores opportunities for research and practice now and going forward focusing on involving people more meaningfully through their experiences and assessments as key for the seemingly intractable challenges facing cities and urban regions globally. Chapter 7 also discusses the use of ambient theory, ambient approaches, and ambient methodologies in smart cities referring to assessments of air quality by people based on their experience of their city. A key takeaway from this chapter is the importance of awareness and the augmenting of awareness in relation to both people and technologies in smart cities.

Chapter 8 provides a synthesis and summary of the preceding seven chapters in developing the interconnectedness of invisibilities and visibilities as InVisibilities. The notion of InVisibilities as ambient in smart environments is addressed, now and for future cities, focusing on associated and emerging research in smart cities. The multiple frameworks developed in chapters 1 to 7 are integrated into a conceptual framework for ambient InVisibilities in smart cities and regions. Observations from the various literature reviews are summarized in a diagram highlighting the elements that emerged by chapter. Relationships for visibilities and invisibilities are presented in tabular form along with the Spearman correlation coefficients that emerged for each. Ideas, insights, and exercise questions are presented in tabular form based on the focus of each chapter. Chapter 8 also provides tabular presentations of future directions as opportunities for research and future directions as opportunities for practice. As a further synthesis, Chapter 8 provides a taxonomy of visibilities and invisibilities in contributing further to the theoretical foundations for smart cities. Reflections are also offered in Chapter 8 along with exercises for the reader.

While this book consists of eight interconnected chapters, each chapter is nevertheless designed to stand on its own and may be read as an independent work. As such, selected elements such as the methodology, limitations, target audience, key terms and definitions, may seem to appear in each chapter when all chapters of the book are viewed and read together. Chapters 1 to 7 present findings; a discussion and interpretation of results; and insights, and ideas. Exercise questions are provided at the end of each chapter as a way for people to become more involved in thinking about particular aspects of smart cities. The exercise format contributes to the creating of an interactive space for sharing in a classroom context where the book or any given chapter may be used as a resource in a classroom setting. Readers are given the option to share their responses to the exercise questions with the author through an online form, available for each chapter.

In summary, key terms constituting the focus of this work include:

- Algorithms
- Ambient Inquiry
- Ambient Theory
- Appreciative Inquiry
- Artificial Intelligence (AI)
- Attuning to Urban Spaces
- Augmented Awareness
- Awareness
- Computational Intelligence and Computer Networking (CI/CN)
- Datawalking
- Design Thinking
- Emotion/Affect
- Heightening Urban Sensibilities
- Imagination
- Information and Communication Technologies (ICTs)
- Intuitive Inquiry and "auspicious bewilderment"
- Invisible Infrastructures
- Place-making
- Self-relevant Research
- Sensors and Sensing
- Translucence
- Transparency
- Urbanizing (e.g., adapting for urban uses)
- Visible Infrastructures

The writing of this book began during the latter part of 2019 and throughout the writing, peer-review, editing, and revising that continued throughout 2020 and into 2021 was the emergence and evolving of the COVID-19 global pandemic. During this time many articles appeared in the popular literature and in research spaces about the pandemic concerning the possible implications, effects, opportunities, and potentials for cities, now and going forward. The Additional Readings sections of some chapters in this book (e.g., Chapters 1, 3, 4, 5, and 6) include items from the popular literature pertaining to the pandemic while Chapter 7 includes discussions of some associated issues such as density, urban edges, health and well-being, inequalities, spatial arrangements, and urban transformation opportunities, to name a few, providing thinking on how cities are being shaped and influenced and how people are rethinking cities and smartness and responding in myriad and creative ways.

Section 1
Visibilities and Invisibilities

Chapter 1
Visibilities and Invisibilities:
Theoretical Perspectives on Smart Cities

ABSTRACT

The purpose of this chapter is to explore the visible and invisible aspects and elements of contemporary urban environments and regions through a review of the research and practice literature on smart cities. Using an exploratory case study approach combined with an explanatory correlational design, the relationship between the visible and invisible is explored with individuals in multiple small- to medium- to large-sized cities in a range of countries. This chapter is significant in that it explores the nature of the relationship between the visible and the invisible in smart urban environments and regions. Key contributions of this work include 1) providing a review of the research and practice literature for urban visibilities and invisibilities in smart cities, 2) exploring theory associated with the visible and invisible in smart cities, and 3) formulating and operationalizing a conceptual framework for visibilities and invisibilities in support of emerging smart city research and opportunities.

1. INTRODUCTION

The purpose of this chapter is to explore the visible and invisible aspects and elements of contemporary urban environments (Dourish and Bell, 2007; Caprotti, 2017; Tursić, 2017; Orasch, 2019) and regions through a review of the research and practice literature on smart cities. To this end, this chapter seeks to identify and describe what contributes to the visible and invisible

DOI: 10.4018/978-1-7998-3850-0.ch001

dimensions of urban spaces and regions generally and more particularly in smart cities and why such dimensions matter. Focus is placed on the physical and digital in relation to the visible and invisible, giving rise to the need for extending and evolving spaces for theory (Brenner and Schmid, 2015; Brenner, 2019) and methodological approaches for understanding and interacting in and with emergent smart city environments that are increasingly being infused with aware technologies. As such, this chapter seeks to shed light on the importance of the relationship between the visible and invisible in urban environments and regions, informed by the experiences and assessments of people from diverse sectors. Using an exploratory case study approach combined with an explanatory correlational design, the relationship between the visible and invisible is explored with people in multiple small to medium to large-sized cities in a range of countries. This chapter is significant in that it explores the nature of the relationship between the visible and the invisible in smart urban environments and regions from the perspectives of people in their day-to-day environments. Key contributions of this work include a) providing a review of the research and practice literature for urban visibilities and invisibilities; b) exploring theory associated with the visible and invisible in smart cities; and c) formulating and operationalizing of a conceptual framework for visibilities and invisibilities in support of emerging smart city research and opportunities.

Objective: The main objective of this chapter is to explore and advance understandings of what contributes to the visible and invisible in smart cities. As such, the key research question posed is – *What is the nature of the relationship between visibilities and invisibilities in smart cities and why does this relationship matter?*

2. BACKGROUND AND OVERVIEW

The notion of the visible and invisible in relation to cities is not new, discussed as it is by Doležel (1979) however, in this work, the emerging context of smart cities provides the novelty. More generally, Malm (2016) speaks of the notion of "visibility impairment" where haze "obscures urban areas." In the United Kingdom, awareness of smart cities is said to remain low with Government Europa (2019) reporting that "68% of UK respondents do not know what a smart city is" or how the smart city would "benefit urban residents." Ransom (2019) argues that "you may already be living in a smart city" adding "here's what to look out for" – highlighting elements such as, more voices, to "promote cooperation between citizens and local institutions"; more complexity, requiring "interdisciplinary work"; and more places for interaction "blurring the edges between the university and the city." Orasch (2019) points to "the inability to visualize" smart cities "and understand the ubiquity of their reach" claiming that "the smart city exists all around us, in various forms and in various spaces." Orasch (2019) goes on to note that many smart city technologies are "invisible to the public eye", contributing to making the visibility of smart urban spaces and regions invisible. Orasch (2019) argues for a "need for public awareness" in the form of education about "what smart city technologies are and where they exist within city space." As such, this chapter elaborates on and extends earlier understandings of urban environments and regions in smart cities in terms of the visible and the invisible (McKenna, 2019). Through a review of the research literature, a context is provided for visibilities and invisibilities along with emerging definitions for key terms used in this chapter. In summary, this chapter explores visibilities and invisibilities through smart city awareness identified in the practice and research literature and explored through ongoing urban research described in this work. This chapter is also significant in that it advances the need for greater awareness of smart cities in terms of the people component (Morello and Piga, 2016) and the experiences of people (Demers and Potvin, 2016) in relation to interactions with aware technologies and with each other.

2.1 Definitions

Definitions are presented here, based on the research literature, for key terms used in this chapter. Additional terms with definitions, as articulated by this work, are included in the *Key Terms and Definitions* section at the end of this chapter.

Invisibilities. According to Brighenti (2017), from a sociological perspective, "the invisible is not the unseen" and "the invisible adds to the visible."

Smart Cities. Urban Hub (2019) claims that "the smart city concept is simply good urban planning" because it "incorporates both advances in digital technology and new thinking in the age-old concepts" from "relationships, community, environmental sustainability" to "good governance, and transparency" adding that, "above all, smart cities are interactive cities."

Visibilities. From a sociological perspective, Brighenti (2017) states that, "visibility concerns an enlarged domain that gathers together and interweaves the sensory and the representational (or, symbolic) registers."

2.2 Overview

This chapter provides a range of perspectives in Section 3 on the visible and the invisible in smart cities based on the research and practice literature, enabling formulation of a conceptual framework for operationalization and use in the exploration under investigation in this chapter. Section 4 provides a description of the methodology used in this chapter followed by the results and a discussion of findings in Section 5. Section 6 identifies opportunities associated with future directions for research and practice and Section 7 concludes with chapter coverage highlights, the major contributions, limitations, and key insights and ideas. Several exercise questions are also included for the reader in support of further exploration.

The primary audiences for this chapter include students, educators, researchers, policy makers, businesses, community members, and urban practitioners concerned with spaces for working with visibilities and invisibilities pertaining to urban design, urban interactions, and associated opportunities in the context of smart cities and future cities and regions.

3. PERSPECTIVES ON THE VISIBLE AND THE INVISIBLE IN SMART CITIES

In the context of cities and smart cities, this chapter provides a review of the research and practice literature for visibilities and invisibilities in relation to the physical and digital and associated complexities pertaining to people and their experiences of urban spaces, regions, and environments.

3.1 Visibilities, Cities, and Smart Cities

Jacobs (1961) described the importance of "a visibly lively public street life" in "accommodating and encouraging, economically and visually, as much variety as possible." Merleau-Ponty (1968) claims that, "the visible is a landscape, a topography yet to be explored." From a human geography and sociological perspective, Shields (2004) uses the notion of "visualicity" as in, “the condition of visibility” to explore urban visibility and invisibility. Lehtovuori (2010) addresses “the problems of the visibility of urban space” arguing that “visibility in the real world can only be attained through the opaqueness of materials.” Regarding the “visible city” Lehtovuori (2010) speaks of the “privileging of the visual” with the “view from above” recalling the work of De Certeau (1993) and the chapter on “walking in the city” describing the view of New York city from the 110th floor of the World Trade Center where “the feeling of ‘seeing the whole’ is enjoyable.” Lehtovuori (2010) claims that “maps and statistics essentially define the ‘visible city’” as in, “the part of urban reality that is taken seriously in the planning and realisation procedures” providing “opportunities to systematically collate data and to make the city visible and intelligible through various theoretical syntheses.” Cairns (2016) describes the notion of a hybrid world encompassing the physical and digital, claiming that, “an invisible digital matrix has become interwoven into the physical infrastructure and sensorial experience of the contemporary city.” This is important because, according to Cairns (2016), “our perception of the city” is altered by this matrix that also “informs our behaviours and increasingly influences the physical designs we propose.” In the context of intelligent technologies, it is worth noting that Caprotti (2017) describes people as embedded in the smart city, as in, being “part of it” in combination with smartphones and other aware technologies. Schmitt (2017) discusses the visible and invisible in relation to “information cities” as flows of information. Brighenti (2017) argues that, “visibility itself can only be understood as a notion of virtuality, not actuality” where the “visible represents a virtual expanse” and drawing on the work of Deleuze (1996), Brighenti (2017) advances a theory of visibility, claiming that “the virtual appears as a ‘mist’ of images surrounding each actual object.” This would seem to describe emerging notions of augmented reality as the use of “technology to superimpose information – sounds, images and text – on the world we see” (Emspak, 2018).

Cowley, Joss, and Dayot (2018) suggest that, “it may be rewarding to focus on the *visibility* of new innovations” where “policy-makers and practitioners” are “actively seeking a more widespread and involved sense of public debate around, and engagement with, the implications of smart technology.” From a law and technology perspective, Brauneis and Goodman (2018) note that “there is no central registry of algorithms in use by governments” adding that, “algorithms are not naturally visible in the way that, say, skyscrapers or bridges are.” Brighenti (2019) addresses measurements in relation to what is rendered visible and invisible. Stephanidis et al. (2019) refer to interactions as being "considerably escalated" in relation to "users' location, posture, emotions, habits, and intentions" as "candidate input data to a variety of visible and invisible technological artifacts embedded in the environment." Fraser (2019) uses art in exploring "representations of the city" drawing on "a selection of comics from across the globe." Wu and Keil (2020) explore the geographies of sub/urban theory focusing on "peripheral areas in fast growing Asian metropolitan regions." Parisi, Holzbach, and Rognoli (2020) note that "classic material categories seem to be increasingly dissolving" such that "technology, materiality, and design are overlapping in the creation of the material provoking and suggesting unique possible scenarios of future practices and applications." Ferrara and Pasetti (2020) introduce the notion of UltraSurfaces (US) described as "electronic augmented material" with "an embedded system of sensors, actuators, smart materials, including communication devices, a microprocessor or edge AI" providing "hybrid materiality" that "enables a seamless fusion of physical matter and digital performance through computational augmentations." Accordingly, Ferrara and Pasetti (2020) claim that, "surfaces acquire intelligence and extraordinary interactive qualities, as well as dynamic behaviour, expressiveness, and personality" and as such, UltraSurfaces "are perceived as *almost-living entities*." In the context of data enabled through smart city technologies Lai, Lai, and Lai (2021) identify the domain of "energy analytics and visibility" showing a "close relationship" with the sub-domains of "sensors and intelligent electronic devices" and "intelligence and data analytics."

3.2 Invisibilities, Cities, and Smart Cities

Merleau-Ponty (1968) argued that "meaning is *invisible*, but the invisible is not the contradictory of the visible" going on to note that "the visible itself has an invisible inner framework (*membrure*), and the in-visible is the secret counterpart of the visible" and "it appears only within it" adding that "one cannot see it there and every effort to *see it there* makes it disappear." However, Merleau-Ponty (1968) claims that the invisible "is *in the line* of the visible" and "it is its virtual focus, it is inscribed within it (in filigree)." Bohn et al. (2005) draw attention to the invisible in terms of technologies embedded in products for a "smart world" that have been "pushed through largely unnoticed by the general public extending quite rapidly into our everyday lives" resulting in smart products and their tracking; anything, anywhere business models; and real time shopping contributing to information asymmetries.

Hollands (2008) critiques the labeling of cities as smart in relation to "the underlying assumptions/contradictions hidden within the concept" looking at "city websites" to see "what things are emphasized and which things are hidden from view." For Harrison and Donnelly (2011), the embedding of technologies in urban areas and regions, together with the unnoticed gathering of large amounts of data, contribute to the notion of smart cities as invisible. Crutzen (2006) speaks in terms of "mental, physical, methodical invisibility" claiming that, "the 'visible' acting of people will be preceded, accompanied and followed by the invisible and visible acting of the AmI technology and their producers" where AmI refers to ambient intelligence. In an interview with the IoT Council (2011), Crutzen describes AmI as, "a technology focused on the 'invisible' interaction with humans" while the Internet of Things (IoT) "is focused on the technological infrastructure for making things interactive and identifiable" and, as such, "is one of the options for realizing AmI." The ambient is described by McCullough (2013) as, "that which surrounds but does not distract" and "a continuum of awareness." Brenner and Schmid (2015) claim that what is going on with non-urban landscapes is rendered invisible by "equating the urban exclusively with large and/or dense population centers." In this way, Brenner and Schmid (2015) highlight a rendering of the invisible made possible through urban discourse, limited by notions of density and the like whereas in an expanded view, the urban encompasses much more, extending beyond traditional boundaries to surrounding regions and territories forming conurbations of economic, cultural, and other activities.

Caprotti (2017) notes that, "the smart city is characterized by its invisibility" in that "fibre networks are material" yet they are not visible because "they are laid on the bottom of canals, or exist underground." Additionally, Caprotti (2017) claims that, "smart city projects are often computer-based and have little or no actual visible, tangible physical presence." In the context of Internet research from the perspective of "connectivity as a public good", Halegoua and Lingel (2017) address hypervisibility and invisibility in urban broadband networks in terms of connectivity and "inactive networks that are buried under the streets" as in "dark fiber" that is "turned off" and "activated networks that are 'turned on' and prominently displayed on city streets" as in "WiFi enabled digital kiosks that provide free internet access to pedestrians drawing attention to the public-ness of infrastructure provision." Halegoua and Lingel (2017) create a space, albeit "contentious", for discussion and debate about "the street" and "the buried potential of dark fiber networks" involving "the constitution of publics, both as civic and internet users."

Tursić (2019) claims that, when "engaged in aesthetic experience, humans exhibit a particular sort of intentionality through which they bring to mind what is not visible through what is present and perceived." Bensai (2020) speculates about the notion of "invisibility cloaks" for the design of smart cities going forward. Shadowen, Lodato, and Loi (2020) address the invisibility of code in Automated Decision Systems (ADS) in smart cities. McCullough (2020) notes that "the formerly invisible world of electricity" is being upended by the several trends of "energy decentralization, decarbonization, Internet of Things, and external threats." Borning, Friedman, and Logler (2020) provide an exploration of information technology in relation to "the largely invisible" material impacts, categorized by the five forces of: disciplinary norms and practices of computer science; metaphors; utopian visions; hardware visibility; and economics. Gardner (2020) reminds us of the "influential essay" by Isozaki Arata in the "Invisible City" where "buildings and cities were momentary fragments in a context of fluidity."

This work is being written during the time of the emerging and evolving of the coronavirus global pandemic and it is worth noting that McCann (2020) uses "four dialectical tensions to understand the COVID-19 city", one of which is "invisibility and visibility", drawing on work by Berman (1982). And finally, Yaylali-Yildiz, Spierings, and Çil (2020) explore publicness as "interaction, discovery, and display" on a university campus where a "visibility score" is

Figure 1. Visibilities and invisibilities: Emerging aspects and domain coverage for smart cities

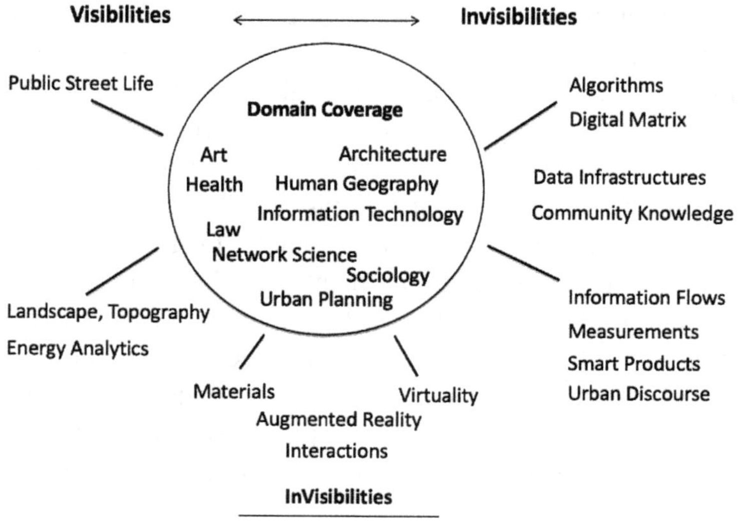

used in analysis and assessments. Yigitcanlar, Kankanamge, and Vella (2021) explore "popular concepts and technologies" using "a systematic Geo-Twitter analysis" to reflect both hidden and dispersed community knowledge about smart cities."

Graphically, emerging aspects of the visible and invisible in smart cities are depicted in Fig. 1 based on this review of the literature, providing a glimpse of the domain coverage, from art to information technology, to urban planning, to name a few.

Fig. 1 also points to the concept of InVisibilities emerging from the overlapping and interweaving of visibilities and invisibilities generally and in the context of smart cities.

3.3 Key Elements Characterizing Visibilities and Invisibilities in Smart Cities

Table 1 provides an overview of key elements characterizing visibilities and invisibilities generally, and in relation to smart cities, emerging from the review of the research literature from 1961 to 2021.

In summary, visibilities encompass public street life (Jacobs, 1961), the landscape and typography to be explored (Merleau-Ponty, 1968), theory (Deleuze, 1996), heuristic concepts such as visualicity (Shields, 2004), what is emphasized (Hollands, 2008), urban space (Lehtovuori, 2010), people (Caprotti, 2017), information (Schmitt, 2017), the virtual (Brighenti, 2017), hypervisibility and invisibility in broadband networks (Halegoua and Lingel, 2017), algorithms (Brauneis and Goodman, 2018), measurements (Brighenti, 2019), innovations (Cowley et al., 2018), augmented reality (Emspak, 2018), global comics (Fraser, 2019), escalated interactions involving data (Stephanidis et al., 2019), geographies of the sub/urban (Wu & Keil, 2020), the dissolving of technology, materiality and design (Parisi et al., 2020), assessments of publicness (Yaylali-Yildiz et al., 2020), energy analytics and visibility (Lai et al., 2021), and "hidden and dispersed community knowledge about smart cities" revealed through a systematic analysis of Geo-Twitter data focusing on how technologies are perceived and utilized (Yigitcanlar et al., 2021).

While invisibilities are said to encompasses in the visible (Merleau-Ponty, 1968), invisibilities in smart cities encompass the mental, physical, and methodical (Crutzen, 2006), the hidden (Hollands, 2008); the ambient with technologies interacting with humans (Crutzen, 2011), embedding of technologies and the unnoticed gathering of data (Harrison & Donnelly, 2011),

Table 1. Overview of key elements characterizing visibilities and invisibilities in smart cities

Characteristics of Visibilities and Invisibilities in Smart Cities			
Author(s)	**Year**	**Visibilities**	**Invisibilities**
Jacobs	1961	"visibly lively public street life"	
Merleau-Ponty	1968	Landscape/typography to explore	In-visible as counterpart of the visible
Deleuze	1996	Theory of visibility	
Shields	2004	Visualicity (heuristic concept)	
Bohn et al.	2005		Smart products and Information asymmetries
Crutzen	2006		Mental, Physical, Methodical
Hollands	2008	What is emphasized	What is hidden
Lehtovuori	2010	Urban spaces	
Crutzen	2011		AmI – Tech & interactions with humans
Harrison & Donnelly	2011		Embedding of tech; unnoticed data collection
McCullough	2013		Ambient – continuum of awareness
Brenner & Schmid	2015		Non-urban landscapes
Cairns	2016		Digital matrix – Physical and sensorial
Caprotti	2017	People as part of intelligent tech	Fibre networks underground/SC tech projects
Schmitt	2017	Information city	
Brighenti	2017	Virtual expanse	
Halegoua & Lingel	2017	Hypervisibility and Invisibility – Urban Broadband Networks	
Brauneis & Goodman	2018	Algorithms	
Cowley et al.	2018	Innovations	
Emspak	2018	Augmented reality	
Brighenti	2019	Measurements	
Fraser	2019	Global comics	
Stephanidis et al.	2019	Escalated interactions/data/etc	
Tursić	2019		What is not present and perceived
Borning et al.	2020		Information technology
Ferrara & Pasetti	2020	UltraSurfaces / hybrid materiality	
Gardner	2020	Liquid cities	
McCullough	2020		Electricity
Parisi et al.	2020	Technology/Materiality/Design	
Shadowen et al.	2020		ADS code
Wu & Keil	2020	Geographies of sub/urban theory	
Yaylali-Yildiz et al.	2020	Assessment of publicness	
Lai et al.	2021	Energy analytics	
Yigitcanlar et al.	2021	Hidden & dispersed community knowledge about SCs – Geo-Twitter data analysis	

ambient awareness (McCullough, 2013), non-urban landscapes (Brenner & Schmid, 2015); digital matrix interweaving physical infrastructure and sensorial experiences in the contemporary city (Cairns, 2016), underground fibre networks and technology-based smart city projects (2017), information (Schmitt, 2017), what is not present and perceived (Tursić, 2019), material impacts of information technology (Borning et al., 2020), electricity (McCullough, 2020), and the code in automated decision systems (ADS) (Shadowen et al., 2020).

Across this vast array of characteristics, intersections emerge for information and measurements, and for the virtual (e.g., virtual expanse and the digital), technology, and data (e.g., algorithms and unnoticed data collection). Also of note are augmented reality and the sensorial. These terms will inform and guide the selection of elements or proxies used in exploring relationships for visibilities and invisibilities in smart cities in this chapter.

3.4 A Framework for Conceptualizing Visibilities and Invisibilities in Smart Cities

Based on the review of the research literature presented in this chapter, an emerging conceptual framework for visibilities and invisibilities in the context of smart cities and regions is advanced, as depicted in Figure 2. Through the combining of the interactive dynamic of aware people and aware technologies, enabling evolving understandings of the nature of smart cities using the constructs of visibilities and invisibilities, dimensions of the physical and digitally-enabled urban areas and regions are explored through emerging perspectives on relationships and on the virtual.

Through emerging understandings of visibilities and invisibilities in urban areas and regions, relationships are explored in this chapter in responding to the research question posed in the introduction of this chapter, and restated as a proposition under exploration in this chapter, as follows:

P: Understanding the nature of the relationship between visibilities and invisibilities is important in smart cities and regions in relation to emergent and evolving spaces for aware people interacting with each other and with aware technologies.

Figure 2. Visibilities and invisibilities: An emerging conceptual framework for smart cities and regions

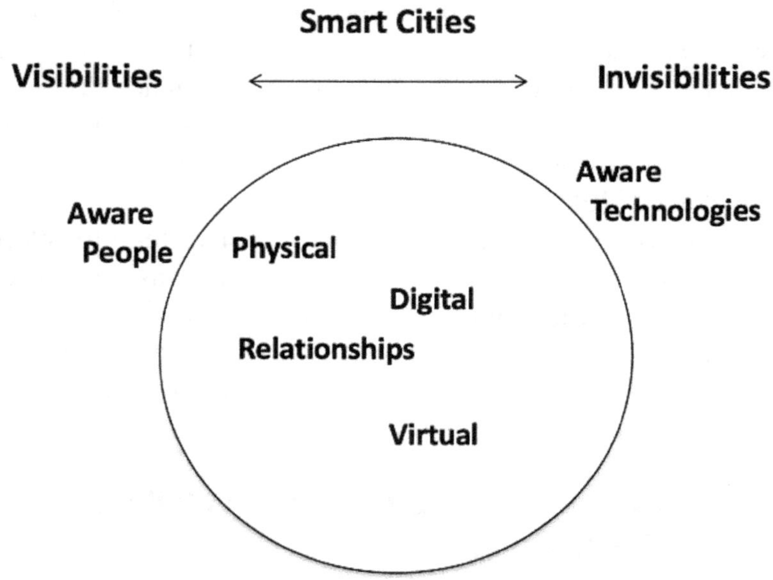

4. EXPLORING VISIBILITIES AND INVISIBILITIES IN SMART CITIES

The methodology for this chapter employs the use of an exploratory case study approach combined with an explanatory correlational design. Yin (2018) points to the "complementarity of case study and statistical research" in that "case studies have been needed to examine the underlying processes that might explain a correlation." People and their experience and assessment of their city as smart in small to medium to large-sized cities in multiple countries forms the case study for this work. According to Yin (2018), the case study approach is particularly appropriate for the study of contemporary phenomena, in this case, smart cities and visibilities and invisibilities. What follows is a description of the methodology for this chapter in terms of the process, data collection, and data analysis.

4.1 Process

A website was used to describe the study and invite participation and demographic data were gathered during sign up including elements such as age range and city name, and people could self-identify in one or more categories (e.g., student, educator, business, community member, city staff, etc.). Once signed up, participants were invited to complete a survey about their experience of the smart city with the opportunity to provide assessments of various urban elements. Participants also were given the option in the survey to be contacted for a more in-depth interview about smart cities. Interviews were conducted in person or online, depending on the location of the participant. Purposive sampling, more specifically heterogeneity sampling, was used as a technique to accommodate a broad spectrum of perspectives (Trochim, 2006). Through an online invitation to participate, interest in the study attracted students, researchers, educators, urban practitioners, and anyone interested in smart cities. Response to the study was received from a diverse range of individuals from cities in Canada, several European countries, Israel, and the United States. Concurrent with this study, group and individual discussions were also conducted systematically, guided by the interview protocol, with a wide range of individuals in multiple Canadian cities.

4.2 Data Collection

Data collection involved the use of an online survey containing 25 questions, consisting of a mix of closed and open-ended. An example of an open-ended survey question is: *What does smartness look like in your city (e.g., smartphones, smart meters, urban displays/screens, sensors, drones, etc.)?* An interview protocol was also used to gather data from in-depth discussions with participants. An example of an interview question is: *What do you like about being in the city?* The survey instrument and the interview protocol were both pre-tested prior to use in the study. As mentioned in the previous section, in parallel with this study, data were also gathered through individual and group discussions in multiple Canadian cities, guided by the interview protocol.

4.3 Data Analysis

Content analysis was used iteratively with qualitative data from interviews, discussions, and open-ended survey questions in the exploration of patterns and relationships. Analysis of quantitative data from survey responses employed the use of descriptive statistics. Using the Real Statistics Software (Zaiontz, 2019) Resource Pack add-in for Microsoft Excel, which contains a variety of statistical functions and data analysis tools, the correlation feature enabled the correlating of items from survey assessments to learn more about the nature of relationships under exploration. Overall, this chapter is based on an analysis of data for n=78 with 41% females and 59% males for people ranging in age from their 20s to their 70s.

5. RESULTS AND DISCUSSION

This section provides an overview of survey, interview, and discussion findings enabling exploration of the relationships between visibilities and invisibilities in smart cities.

5.1 Results for Explorations of Visibilities and Invisibilities in Smart Cities

In the survey for this study, participants were asked to assess to what extent "city-focused social media and other aware technologies give rise to many possibilities" such as *attuning to urban spaces* using a 7-point Likert-type scale (e.g., 1 = Not at all, 2 = Not sure, 3 = Maybe, 4 = Neutral, 5 = Sort of, 6 = Sure, 7 = Absolutely). Table 2 shows the results of the survey assessment for *attuning to urban spaces* aided by social media and aware technologies with 17% of responses at position 4 on the scale, 33% of responses at position 6 on the scale and 50% of responses at position 7. In the survey for this study, participants were asked to "assess to what extent elements such as *visible infrastructures* contribute to the making of smart cities, using a 7-point Likert-type scale (e.g., 1 = Not at all, 2 = Not sure, 3 = Maybe, 4 = Neutral, 5 = Sort of, 6 = Sure, 7 = Absolutely). Results for visible infrastructures show 25% of responses at position 5 and 75% of responses at position 7.

Table 2. Assessments for attuning to urban spaces and visible infrastructures in smart cities

Items	Assessments	Correlation
Attuning to urban spaces (social media & Aware tech)	17% (4); 33% (6); 50% (7)	.29
Visible Infrastructures	25% (5); 75% (7)	

Using the correlation feature of the Real Statistics Software (Zaiontz, 2019) in Microsoft Excel, *attuning to urban spaces* aided by social media and aware technologies is correlated with *visible infrastructures,* showing a Spearman correlation coefficient for ordinal data of .29.

Table 3 shows the results of the assessment for *attuning to urban spaces* when correlated with *invisible infrastructures* showing 25% of responses at position 5; 50% at position 6; and 25% of responses at position 7.

Table 3. Assessments for attuning to urban spaces and invisible infrastructures in smart cities

Items	Assessments	Correlation
Attuning to urban spaces (social media & aware tech)	17% (4); 33% (6); 50% (7)	.82
Invisible Infrastructures	25% (5); 50% (6); 25% (7)	

Using the correlation feature of the Real Statistics Software (Zaiontz, 2019) in Microsoft Excel, *attuning to urban spaces* is correlated with *invisible infrastructures*, showing a Spearman correlation coefficient for ordinal data of .82.

Exploring further, using information and communications technologies (ICTs) to represent understandings of the invisible, a correlation was conducted between *ICTs* and *visible infrastructures.* Table 4 shows the results of the assessment for *ICTs* with 50% of responses at position 5 on the scale and 50% of responses at position 7. Results for *visible infrastructures* show 25% of responses at position 5 and 75% of responses at position 7. When assessments of *ICTs* as a proxy for *invisible infrastructures* were correlated with *visible infrastructure* assessments, a Spearman correlation of .57 resulted.

Looking at the scatterplot for this relationship, as ICTs decrease, level out, or increase, so too do visible infrastructures, as depicted in Fig. 3.

Qualitative data gathered from interviews, individual and group discussions, and open-ended survey questions were also analyzed to provide additional insight into the nature of the relationship between visibilities and invisibilities in smart cities. When asked, *"What enables you to feel the pulse of the city?"* an educator responded by observing that, "on the physical level, the number of people who are out and about on a sunny weekend, in the evening" and "on the digital level, keeping up-to-date with social media and online newspapers

Table 4. Assessments for ICTs (invisible infrastructures) and visible infrastructures in smart cities

Items	Assessments	Correlation
ICTs (invisible infrastructures)	50% (5); 50% (7)	.57
Visible Infrastructures	25% (5); 75% (7)	

Figure 3. Scatterplot for ICTs visible infrastructures in smart cities

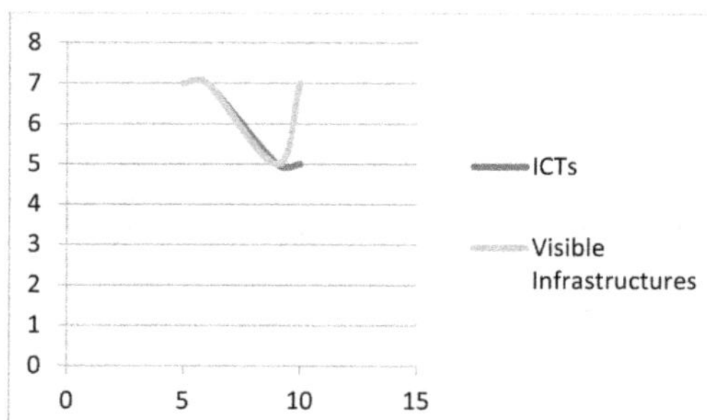

allows me to find out things that my eyes could not see while walking around" adding that, "its' all part of the 'pulse of the city'."

When asked, "*What other key elements contribute to your idea of a smart city?*" a community leader/member pointed to the need for "finding fun and meaningful ways to engage the public" as well as to "visualize this data" so as to "simplify it for the larger public." A community member highlighted the importance of what cities do to create "the visual sense of the city" while a city councillor spoke in terms of "the vibrancy of the city."

When asked, "*What does smartness look like in your city?*" an educator responded, "nothing visible at all, but on a smart device you can see a lot of wifi availability." This individual added that, "no one needs (or maybe wants) a lot of visible infrastructure unless it is 'awesome'" as in "think beautiful data visualizations in real time" and "synonymous with social, useful, and having clear purpose." A community leader/member commented that, "smartness in this city is increasing from a smart phone perspective and apps available, but we do not see too many physical manifestations of this yet" although "I just noticed a bike counter the other day down by the Railyards" acknowledging that "perhaps I just haven't encountered them!" An individual from the business sector stated, "not sure I see much IoT stuff, except an advanced parking system using the Pango app, and other transportation innovations."

The notion of visible and invisible infrastructures emerged in the form of an urban display to dynamically capture and show cycling data on a visual screen, in-the-moment, for a multi-use path. A city councilor referred to BikeMaps.org as "a great example" where "through your phone you are able to track accidents" and "maybe find the hub, find the place where accidents are happening and with the app create a map, and with that information" as "a real-life thing" via "a phone app, that information is being given back to the city so we can now go and create engineering fixes for those intersections and those high crash places."

Visibility and invisibility were described by information technology (IT) staff in a university setting in the form of occupancy data in building systems on campus, posing the question "how do you leverage and expose some of the information in a way that you know is safe" in order to "make it available for other things" such as presence information to determine whether there are "people in the building and if there is a response" in the case of an emergency situation. A city councilor highlighted the use of technology in an e-TownHall meeting enabling physical and digital attendance for the public through the use of real-time streaming where people, in turn, use social media to comment and provide feedback, generating "documented engagement." City IT staff commented that, "we are interested in putting physical infrastructure in place" adding that, "how we interpret using it is still open." Regarding data, City IT staff indicated that, "we're starting to look at the tools that would give us the visualization" while acknowledging, "we're very much immature in that overall data sense." City IT staff referred to the notion of "data analysis that you've not thought of" as in, "serendipitous or accidental" or "unintended" when tools are made available to collect data that were not "even in our mindset" yielding possibly new forms of value.

5.2 A Discussion and Interpretation of Findings

The presence of a positive correlation between *attuning to urban spaces* (aided by social media and aware technologies) and *visible infrastructures* was achieved in this work (.29). However, Creswell (2018) cautions that a correlation in the .20 to .35 range is considered to have "only a slight relationship" and "may be valuable to explore the interconnection of variables but of little value in prediction studies." A positive correlation was also achieved between *attuning to urban spaces* and *invisible infrastructures* at .82, in the context of smart cities. Creswell (2018) advises that such a correlation "would be considered very good" and that "good prediction can result from one variable to the other." Attuning is suggestive of people and their feeling and sensing capabilities in urban environments as a possible urban metric or indicator. Exploring further using *information and communication technologies (ICTs)* as a proxy for *invisible infrastructures*, a correlation of .57 emerges in relation to *visible infrastructures*. According to Creswell (2018), correlations in the range of .35 to .65 "are useful for limited prediction." Also, a correlation in this range is said to provide "values used to identify variable membership in the statistical procedure of factor analysis" for the "intercorrelation of variables" (Creswell, 2018) and this could be helpful for informing future smart city initiatives, projects, and interventions.

For the city councilor, the example of the BikeMaps.org initiative provided an example of the coming together of visible and invisible infrastructures, such that, "it is all connected and the tool of technology is augmenting the information that we have and allows us to have a safer more vibrant city." It is worth noting that the physical and digital response by an educator encompasses both the visible (physical) and invisible (digital) dimensions, in support possibly of the notion by Merleau-Ponty (1968) that "the in-visible is the secret counterpart of the visible" when experienced in the context of smart cities and regions.

The example provided by the city councilor of an eTownHall meeting highlights how the use of technology enables people, as well as city hall as an organization, to extend notions of visibility through social media and other e-based approaches to meetings and processes. Notwithstanding the need for increased education and funding, City IT staff acknowledged the potential for data visualizations in support of increased value and use in urban areas and regions. As such, examples emerging from the qualitative findings begin to shed light on why the relationships between visibilities and invisibilities matters in smart cities and regions, the nature of such relationships, and possible types of variables for consideration.

6. EMERGING RESEARCH AND OPPORTUNITIES

Opportunities for research and practice emerging from this chapter are identified in sections 6.1 and 6.2 based on the literature review and the findings highlighted from this study.

6.1 Research Opportunities

Research opportunities associated with visible and invisible dimensions of smart cities are perhaps as abundant as they are varied.

Invisible Infrastructures. Research opportunities associated with invisible infrastructures range from those pertaining to urban data visualizations, to sensing and attuning, to urban activities aided by smartphone apps for biking, parking, and food, to name a few.

Visible Infrastructures. Learning more about the capabilities and potential of existing and emerging urban elements such as urban displays provides opportunities yielding information beyond bike count information. For example, does the display infrastructure also track pedestrian counts and could urban displays be enhanced to become more interactive as suggested by an individual from industry.

Opportunities also exist for further testing and validation of the viability of the framework advanced in this chapter for urban visibilities and invisibilities in smart cities.

6.2 Opportunities for Practice

Practice opportunities associated with the visible and invisible dimensions of smart cities that emerged from this chapter are highlighted below in terms of the physical and digital.

Physical. The interweaving of the physical and digital such that digital dimensions can be added to physical infrastructures such as urban displays to enable increased interactions.

Digital. Conceptualization of spaces for real-time, dynamic, and adaptive types of approaches to understanding the invisible and visible emerge in this chapter opening the way for new techniques and possibly novel roles and collaborations for citizens, businesses, researchers, and governments. As such, areas for exploration potential include:

1. *Technology infrastructures* fostered through urban initiatives focusing on animating public spaces with dynamic and interactive displays and visualizations
2. *Human infrastructures* fostered through the making of more dynamic urban spaces utilizing social media and other tools designed to enhance vibrancy, the sharing of experiences, and the pulse of the city

Additionally, this work directs attention to the potential for understanding more about the people dimension of smart cities in terms of more aware people interacting and engaging with more aware technologies and with each other.

7. CONCLUSION

This chapter provides an overview of smart cities and regions in relation to what is rendered visible and invisible from a range of disciplinary perspectives spanning urban planning to architecture. The main contributions of this chapter include: a) development and operationalization of a conceptual framework for visibilities and invisibilities in the context of smart cities; b) a review of the research literature for visibilities and invisibilities; and c) an invitation for the re-thinking of existing urban interpretations taking into consideration the intangible in relation to emerging technologies calling into play increased awareness in fostering smarter, more responsive, cities and regions.

Limitations of this chapter pertain to the small sample size of the study upon which this exploration is based. Additionally, the seemingly abstract and intangible nature of invisibilities in smart cities is mitigated by actual examples from the research literature and in-depth discussions with a range of individuals regarding everyday emergent elements such as data visualizations. *Visible infrastructures* and *invisible infrastructures* are explored in relation to people *attuning to urban spaces* aided by social media and other aware technologies, where correlations of .29 and .82 were found, respectively. It is worth noting that a correlation of .57 was found between *visible infrastructures* and the use of *ICTs* as a proxy for *invisible infrastructures*. As such, potential limitations related to the use of variables are mitigated by the potential to further define and refine understandings of visibilities and invisibilities, going forward.

This chapter provides an exploration of the invisibilities and visibilities constructs in relation to assessments of the experience of urban areas and regions as smart cities. Findings in this chapter assist in identifying the nature of the relationships between *invisibilities* and *visibilities* in urban areas and regions and why such relationships matter for smart cities, highlighting key insights and ideas, while offering exercises for the reader in support of engagement and further exploration.

Insights of particular interest to the reader include the following:

1. The presence of a promising relationship between *ICTs* as a proxy for *invisibilities* when correlated with *visibilities* in evolving understandings of smart cities
2. The interweaving of the visible and the invisible in smart urban environments and regions
3. The notion of unforeseen value emerges in relation to the serendipitous generation, collection, and use of urban public data as a type of invisibility with latent visibility potential

Ideas emerging from this chapter of interest to the reader include the following:

1. The opportunities potential that invisible infrastructures as ICTs present in smart cities and regions
2. The potential for people to engage more meaningfully with invisible spaces in digital/physical urban spaces
3. The opportunities potential for visibly and invisibly interacting with infrastructures in smart cities and regions

Exercises for the reader to consider, encouraging further exploration include the following:

1. Identify one or more smart elements in your city that are visible to the public
2. Identify one or more smart elements in your city that are invisible to the public
3. Describe one or more ways in which smart elements in your city that are invisible to the public could be meaningfully interacted with

If you would like to share your responses to the Exercises with the author, please complete the questions in this online space for Chapter 1 – Visibilities and Invisibilities

This work will be of interest to a broad audience including students, educators, researchers, artists, urban planners, urban practitioners, community members, and anyone concerned with visibilities and invisibilities and associated research and practice opportunities in smart, responsive, and future cities.

REFERENCES

Bensai, R. (2020). Microwave surfing: Roman invisibility cloak? *IEEE Microwave Magazine, 14*, 118.

Berman, M. (1982). *All that is solid melts into air: The experience of modernity*. Simon & Schuster.

Bohn, J., Coroamă, V., Langheinrich, M., Mattern, F., & Rohs, M. (2005). Social, economic, and ethical implications of ambient intelligence and ubiquitous computing. In W. Weber, J. M. Rabaey, & E. Aarts (Eds.), *Ambient Intelligence*. Springer. doi:10.1007/3-540-27139-2_2

Borning, A., Friedman, B., & Logler, N. (2020). The 'invisible' materiality of information technology. *Communications of the ACM, 63*(6), 57–64. doi:10.1145/3360647

Brauneis, R., & Goodman, E. P. (2018). Algorithmic transparency for the smart city. *The Yale Journal of Law of Technology*, *20*, 103–176.

Brenner, N. (2019). *New urban spaces: Urban theory and scale question*. Oxford University Press. doi:10.1093/oso/9780190627188.001.0001

Brenner, N., & Schmid, C. (2015). Towards a new epistemology of the urban? *City*, *19*(2-3), 151–182. doi:10.1080/13604813.2015.1014712

Brighenti, A. M. (2017). The visible: Element of the social. *Frontiers in Sociology*, *2*(17), 1–11. doi:10.3389/fsoc.2017.00017

Brighenti, A. M. (2019). Umwelt-measures. On extensive and intensive measures: Introduction to the special issue 'Theorising measures, rankings and metrics'. *Social Sciences Information. Information Sur les Sciences Sociales*, *58*(2), 224–237. doi:10.1177/0539018419858816

Cairns, G. (2016). Foreword. In G. A. Caldwell, C. H. Smith, & E. M. Clift (Eds.), *Digital Futures and the City of Today: New Technologies and Physical Spaces. Mediated Cities Series*. Intellect Ltd. doi:10.1016/B978-0-12-802032-6.06001-1

Caprotti, F. (2017). Research in the invisible city: Challenges for 'knowing' in the smart city. *UGEC Viewpoints: A Blog on Urbanization and Global Environmental Change*. Retrieved 17 June 2018 from https://ugecviewpoints.wordpress.com/2017/04/20/research-in-the-invisible-city-challenges-for-knowing-the-smart-city/

Cowley, J., Joss, S., & Dayot, Y. (2018). The smart city and its publics: Insights from across six UK cities. *Urban Research & Practice*, *11*(1), 53–77. doi:10.1080/17535069.2017.1293150

Creswell, J. W. (2018). *Educational Research: Planning, Conducting, and Evaluating Quantitative and Qualitative Research* (6th ed.). Pearson.

Crutzen, C. K. M. (2006). Invisibility and the meaning of ambient intelligence. *International Journal of Information Ethics*, *6*, 52–62.

De Certeau, M. (1984). *The practice of everyday life*. University of California Press.

Deleuze, G. (1996). L'actuel et le virtuel. In Dialogues. Paris: Flammarion.

Demers, C., & Potvin, A. (2016). From history to architectural imagination: A physical ambiences laboratory to interpret past sensory experiences and speculate on future spaces. *Ambiances: International Journal of Sensory Environment, Architecture and Space*, 756.

Doležel, L. (1979). The visible and the invisible Petersburg. *Russian Literature*, *7*(5), 465–490. doi:10.1016/0304-3479(79)90008-5

Dourish, P., & Bell, G. (2007). The infrastructure of experience and the experience of infrastructure: Meaning and structure in everyday encounters with space. *Environment and Planning. B, Planning & Design*, *34*(3), 414–443. doi:10.1068/b32035t

Emspak, J. (2018). *What is augmented reality?* LiveScience. Retrieved 1 January 2020 from https://www.livescience.com/34843-augmented-reality.html

Ferrara, M., & Pasetti, C. (2020). UltraSurfaces: A New Material Design Vision. In T. Ahram, W. Karwowski, A. Vergnano, F. Leali, & R. Taiar (Eds.), *Intelligent Human Systems Integration 2020. IHSI 2020. Advances in Intelligent Systems and Computing* (Vol. 1131). Springer.

Fraser, B. (2019). *Visible cities, global comics: Urban images and spatial form*. University Press of Mississippi.

Gardner, W. O. (2020). Liquid cities. *Places Journal*. Retrieved 22 June 2020 from https://placesjournal.org/article/liquid-cities/

Government Europa. (2019). *Smart city investment report: UK awareness remains low*. Government Europa. Retrieved 23 December 2019 from https://www.governmenteuropa.eu/smart-city-investment-report/92975/

Halegoua, G., & Lingel, J. (2017). Invisibility and hypervisibility: Imagining publics for urban broadband networks. *The 18th Annual Conference of the Association of Internet Researchers.*

Harrison, C., & Donnelly, I. A. (2011). A theory of smart cities. *Proceedings of the 55th Annual Meeting of the International Society for the Systems Sciences*. University of Hull Business School.

Hollands, R. G. (2008). Will the real smart city please stand up? Intelligent, progressive or entrepreneurial? *City*, *12*(3), 303–320. doi:10.1080/13604810802479126

Hub, U. (2019). *Cities. Essen, Germany: Thyssenkrupp*. Retrieved 28 December 2019 from https://www.urban-hub.com/cities/

IoT Council. (2011). *Council interview with Cecile Crutzen on ambient intelligence and Internet of Things.* Retrieved 23 January 2020 from https://www.theinternetofthings.eu/content/council-interview-cecile-crutzen-ambient-intelligence-and-internet-things

Jacobs, J. (1961). *The death and life of great American cities*. Random House.

Lai, C. S., Lai, L. L., & Lai, Q. H. (2021). Smart city. In *Smart Grids and Big Data Analytics for Smart Cities* (pp. 1–171). Springer. doi:10.1007/978-3-030-52155-4_1

Lehtovuori, P. (2010). *Experience and conflict: The production of urban space*. Ashgate Publishing.

Malm, W. C. (2016). *Visibility: The seeing of near and distant landscape features*. Elsevier.

McCann, E. (2020). Listening to the city in a global pandemic. In What's it like to live in a city as a global pandemic takes hold? *CityRoadPodcast.* Retrieved 23 May 2020 from https://cityroadpod.org/2020/03/29/listening-to-the-city-in-a-global-pandemic/

McCullough, M. (2013). *Ambient commons: Attention in the age of embodied information*. The MIT Press. doi:10.7551/mitpress/8947.001.0001

McCullough, M. (2020). *Downtime on the microgrid: Architecture, electricity, and smart city islands*. MIT Press. doi:10.7551/mitpress/11953.001.0001

McKenna, H. P. (2019). Visibilities and invisibilities: Methodologies and theoretical spaces for ambient urbanities. In *H. P. McKenna, Ambient Urbanities as the Intersection between the IoT and the IoP in Smart Cities* (pp. 192–222). IGI Global. doi:10.4018/978-1-5225-7882-6.ch007

Merleau-Ponty, M. (1968). *The visible and the invisible* (C. Lafort, Ed., LingisA., Trans.). Northwestern University Studies in Phenomenology & Essential Philosophy.

Morello, E., & Piga, B. E. A. (2016). Experiential simulation in architecture and urban space: Introduction to the special issue. *Ambiances: International Journal of Sensory Environment, Architecture and Space*, 671.

Orasch, A. K. (2019). Seeing the smart city: Mapping technologies in Canada. *The Conversation*. Retrieved 19 December 2019 from https://theconversation.com/seeing-the-smart-city-mapping-technologies-in-canada-112458

Parisi, S., Holzbach, M., & Rognoli, V. (2020). The Hybrid Dimension of Material Design: Two Case Studies of a Do-It-Yourself Approach for the Development of Interactive, Connected, and Smart Materials. In T. Ahram, W. Karwowski, A. Vergnano, F. Leali, & R. Taiar (Eds.), *Intelligent Human Systems Integration 2020. IHSI 2020. Advances in Intelligent Systems and Computing* (Vol. 1131). Springer. doi:10.1007/978-3-030-39512-4_140

Ransom, J. (2019). New wave of smart cities has arrived - and they're nothing like science fiction *The Conversation*. Retrieved 24 December 2019 from https://theconversation.com/new-wave-of-smart-cities-has-arrived-and-theyre-nothing-like-science-fiction-119855

Schmitt, G. (2017). *Future Cities*. MOOC (Massive Open Online Course). Zurich, Switzerland: ETHzurich. Retrieved 11 December 2017 from https://www.edx.org/course/future-cities-ethx-fc-01x

Shadowen, N., Lodato, T., & Loi, D. (2020). Participatory governance in smart cities: Future scenarios and opportunities. *Proceedings of HCII2020*, 443-463. 10.1007/978-3-030-50344-4_32

Shields, R. (2004). Visualicity – on urban visibility and invisibility. *Visual Culture in Britain*, *15*(1), 23–36.

Stephanidis, C., Salvendy, G., Antona, M., Chen, J. Y. C., Dong, J., Duffy, V. G., Fang, X., Fidopiastis, C., Fragomeni, G., Fu, L. P., Guo, Y., Harris, D., Ioannou, A., Jeong, K. K., Konomi, S., Krömker, H., Kurosu, M., Lewis, J. R., Marcus, A., ... Zhou, J. (2019). Seven HCI Grand Challenges. *International Journal of Human-Computer Interaction*, *35*(14), 1229–1269. doi:10.1080/10447318.2019.1619259

Trochim, W. M. K. (2006). *Research Methods Knowledge Base.* Available online: http://www.socialresearchmethods.net/kb/sampnon.php

Tursić, M. (2017). *Aesthetic space: The visible and the invisible in urban agency* (Thèse No. 6445). Lausanne, Switzerland: EPFL.

Tursić, M. (2019). *Aesthetic space: The visible and the invisible in urban agency* (Thèse No. 6445). Lausanne, Switzerland: EPFL. Retrieved 27 January 2020 from https://mirzatursic.com/science-means-constantly-walking-between-influence-and-openness/

Wu, F., & Keil, R. (2020). Changing the geographies of sub/urban theory: Asian perspectives. *Urban Geography*, *41*(7), 947–953. Advance online publication. doi:10.1080/02723638.2020.1712115

Yaylali-Yildiz, B., Spierings, B., & Çil, E. (2020). *The spatial configuration and publicness of the university campus: Interaction, discovery, and display on De Uithof in Utrecht.* Urban Des Int. doi:10.105741289-020-00130-w

Yigitcanlar, T., Kankanamge, N., & Vella, K. (2021). How are smart city concepts and technologies perceived and utilized? A systematic geo-twitter analysis of smart cities in Australia. *Journal of Urban Technology*, *28*(1-2), 135–154. doi:10.1080/10630732.2020.1753483

Yin, R. K. (2018). *Case study research and applications: Design and methods* (6th ed.). Sage.

Zaiontz, C. (2019). *Real Statistics using Excel.* www.real-statistics.com

ADDITIONAL READING

Caprotti, F. (2019). Spaces of visibility in the smart city: Flagship urban spaces and the smart city imaginary. *Urban Studies (Edinburgh, Scotland)*, *56*(12), 2465–2479. doi:10.1177/0042098018798597

Cheng, D. (2013). (In)visible urban water networks: The politics of non-payment in Manila's low-income communities. *Environment and Urbanization*, *25*(1), 249–260. doi:10.1177/0956247812469926

Constable, H. (2020). How do you build a city for a pandemic? *BBC Future*. Retrieved 1 May 2020 from https://www.bbc.com/future/article/20200424-how-do-you-build-a-city-for-a-pandemic

Fortune (2020). Special report: Rethinking the city. *Fortune*. Retrieved 28 April 2020 from https://fortune.com/packages/special-report-rethinking-the-city/

Marchessault, J., & Lord, S. (2016). *The visible city project + archive*. Toronto: York University. Retrieved 11 June from https://visiblecity.info.yorku.ca/project/

Ribarsky, W., Wasilewski, T., & Faust, N. (2002, July-August). From urban terrain models to visible cities. *IEEE Computer Graphics and Applications*, *22*(4), 10–15. doi:10.1109/MCG.2002.1016692

Serhan, Y. (2020). Vilnius shows how the pandemic is already remaking cities: Outbreaks of disease have shaped urban life for centuries. *Atlantic (Boston, Mass.)*, *9*(June). Retrieved June 9, 2020, from https://www.theatlantic.com/international/archive/2020/06/coronavirus-pandemic-urban-suburbs-cities/612760/

Wall, K. (2018). Visible cities. Poems by Kathleen Wall and Photographs by Veronica Geminder. Calgary, Alberta: University of Calgary Press. doi:10.2307/j.ctt2121799

KEY TERMS AND DEFINITIONS

Ambient: In the context of smart cities, the ambient describes the increasing presence of more aware people together with awareness-enabled technologies in support of interactivity, influencing human communication and activities, human-machine interactivities, and machine-machine interactivities and in turn affecting behavior, experience, information capabilities, economies, literacies, and indeed everything.

Augmentations: Augmentations refer to the extending of capabilities involving the use of technologies, as in for example, augmented reality and the visual overlay of information in everyday spaces, contributing to heightened forms of awareness.

Awareness: Awareness is conceptualized in this work as the quality of being aware as it applies to people on the one hand, to the enabling of technologies on the other, and to the enhancing of awareness through people interacting with awareness-enabled and enabling technologies.

Invisibilities: Invisibilities refer to elements that are rendered invisible or that may present as intangibles.

Smart Cities: Smart cities are urban areas and regions characterized by more aware and engaged people, interacting with each other and aided by the use of more awareness enhancing technologies for a wide range of purposes from communication to design to atmospheres to mobility to livability, to governance, to data generation and use for learning and informed decision-making.

Visibilities: Visibilities refer to physical and other visible elements and infrastructures.

Chapter 2
Making the Invisible Visible:
Practical Perspectives in Smart Cities

ABSTRACT

The purpose of this chapter is to explore how the invisible is being made visible in smart cities and regions. The research literature for urban approaches to making the invisible visible is explored in this chapter in the context of smart and responsive cities and regions, enabling the identification of opportunities for research and practice. Using an exploratory case study approach combined with an explanatory correlational design, this chapter reveals how people describe and assess their experience of cities as smart. An analysis of quantitative and qualitative data focusing on the constructs of heightening urban sensibilities and urbanizing sheds light on opportunities for both practice and research going forward. This chapter makes a contribution to 1) the research literature for urban approaches to the making of smarter and more responsive cities through making the invisible visible and 2) conceptual and practical thinking through formulation and operationalization of an emerging framework for making the invisible visible in urban spaces and regions.

1. INTRODUCTION

The invisible has been articulated by researchers in the context of cities and urban areas and regions in terms of the everyday use of infrastructures (Dourish and Bell, 2007), information flows (Schmitt, 2017), the instrumented

DOI: 10.4018/978-1-7998-3850-0.ch002

city and associated discourse (Mattern, 2016), and novel data from a human geography perspective, influencing social processes and experiences (Bian, 2020), to name a few. In exploring how the invisible is being made visible in the context of smart cities and regions, this chapter seeks to identify practical examples in everyday life. As such, this chapter explores how the invisible is being made visible in smart urban environments and regions based on a review of the practice and research literature on the one hand and through exploring with people their experience of the city as smart on the other, using a case study approach. An explanatory correlational design is also used in this chapter to explore the relationship between urban elements such as *heightening urban sensibilities* and *urbanizing*, as in, *adapting for urban uses*. This type of correlation could contribute to greater understandings of the nature of elements and indicators contributing to smart cities and to the success potential for initiatives, interventions, and projects.

Objective: The objective of this chapter is to explore and advance opportunities for people to engage in making the invisible visible in smart urban environments and regions. As such, the key research question posed is – *What does making the invisible visible mean in smart cities and why does this activity matter*?

2. BACKGROUND AND OVERVIEW

Hill (2008) observes that "the invisible becomes visible, as the impact of people on their urban environment can be understood in real time" using the example of "water use patterns" and the turning on of taps. Researchers have articulated the invisible dimensions of cities and smart cities from multiple perspectives (Rickert, 2013; McCullough, 2020; Schmitt, 2015; Orasch, 2019) with Caprotti (2017) claiming invisibility to be a characteristic of smart cities. In defining smart cities, Sáinz-Piña (2011) points to "a city that uses information and communication technologies to make its critical infrastructure, its components, and public services more interactive, efficient and visible to citizens." Mattern (2014) argues for the importance of considering "the point of engagement where citizens interface with the city's operating system" noting that, "the zone between person and machine is only the most visible type of interface." Mattern claims that urban dashboards (2015) "render a city's infrastructures visible and make tangible, or in some way comprehensible, various hard-to-grasp aspects of urban quality-of-life." Schmitt (2015) highlights the importance of information architecture in that "it visualises the information inherent in a building and thus makes the invisible visible" in the context of "the built environment" in smart cities.

How invisible dimensions of smart cities are being made visible is the focus of this chapter, drawing on insights from the practice and research literature (Caprotti, 2018; Chevalier, 2018; Servières et al., 2017). While intelligent, aware, and increasingly pervasive technologies contribute to making the invisible visible in smart cities (Schmitt, 2017) other elements are also present in the form of how people are interacting with information infrastructures (Dourish and Bell, 2007) and with smart urban initiatives (Mattern, 2020), contributing to emergent opportunities and potentials. While the notion of the invisible in relation to cities is not a new one, addressed as it is in works such as the novel by Calvino (1972; 1974), what is different about this chapter is the notion of making the invisible visible in the context of emerging understandings of smart cities and regions, and the notion of urban visibilities and invisibilities in this work more broadly. As such, this chapter seeks to learn more about people interacting with and experiencing their urban environments and spaces in coming to more dynamic understandings of smart cities and regions. It is worth noting that collaboration figures strongly among ideas for addressing the challenges of global cities (Belanger and Hinchliffe, 2020), as part of a special report on "rethinking the city."

2.1 Definitions

Definitions are presented here, based on the research literature, for key terms used in this chapter. Additional terms with definitions, as articulated by this work, are included in the *Key Terms and Definitions* section at the end of this chapter.

Augmentations. Servières et al. (2017) refer to "an augmented urban experience" in the context of "connected wearable tools" contributing to "a change in everyday life."

Emotion/Affect. Feldman Barrett, Mesquita, Ochsner, and Gross (2007) note, "this view that an experience of emotion is a state of mind whose content is at once affective (pleasant or unpleasant) and conceptual (a representation of your relation to the world around you) – is consistent with recent theoretical insights in the neurobiology of consciousness."

Information Architecture. Schmitt (2015) defines information architecture "as the necessary framework to understand architecture, urban systems, and territories in the knowledge society."

Urban Displays. Shilov et al. (2019) note that urban displays "can be a very useful information source in smart cities since they can be observed by a large amount of people." Citing several researchers, such displays are referred to as "digital signage" (Schaeffler, 2008; Want and Schilit, 2012) described as "the provision of content (advertisements, news, assistance) on electronic large displays in places where many people are present or passing by" (Wißotzki, 2017).

Visualizations. Moore (2017) points to the importance of visualization for enabling "aggregated and integrated data from various sources" so that it "can be 'seen' in a way that produces actionable intelligence."

2.2 Overview

The notion of ambient sensing was advanced by McKenna (2019), calling for explorations into questions such as "how are multi-sensorial capabilities affected, augmented, or extended by the use of say, smartphones, other smart devices, and sensing applications in urban spaces?" and "how are experiences of sensing the pulse, vibrancy, space, and spatialities of the city affected (positively or adversely) by the use of smartphones, sensing apps, and other sensing technologies?" This work considers sensing that is technology based and human based, augmented in both cases by awareness capabilities. A range of perspectives are provided in Section 3 on making the invisible visible in smart cities based on the research and practice literature, enabling formulation of a conceptual framework for operationalization and use in the explorations conducted in this chapter. Section 4 provides a description of the methodology used in this work, followed by the results, and a discussion of findings. Section 5 identifies opportunities associated with future directions for research and practice and Section 6 concludes with chapter coverage highlights, the major contributions, limitations, and key insights and ideas. Several exercise questions are also included for the reader in support of further exploration.

The primary audiences for this chapter include students, artists, engineers, educators, researchers, policy makers, businesses, community members, and urban practitioners concerned with approaches for working with making the invisible visible, associated with urban design, urban representation, urban interactions, and other challenges and opportunities in the context of smart cities and regions and future cities.

3. PERSPECTIVES ON MAKING THE INVISIBLE VISIBLE IN SMART CITIES

In the context of smart cities, this chapter provides a review of the research and practice literature for approaches to making visible the invisible; key approaches in relation to the physical, digital, and less tangible; and associated complexities pertaining to the urban for people and their experiences.

3.1 Moving From the Invisible to the Visible in Smart Cities

Approaches to making the invisible visible in smart cities are considered in relation to computational intelligence (CI) and communication networks (CN), emotion/affect, imagination, infrastructure, information, simulations, electricity, discourse, art and visualizations, technology, and human geography.

Computational Intelligence (CI) and Communication Networks (CN)

Jordan (2019) identifies "significant misunderstanding accompanying use of the phrase" artificial intelligence (AI) and this "distracts us" since, in large part, "what is labeled AI today, particularly in the public sphere, is actually machine learning (ML)" described as "an algorithmic field that blends ideas from statistics, computer science and many other disciplines to design algorithms that process data, make predictions, and help make decision." Jordan (2019) notes that "not visible to the general public" considerable work is occurring behind the scenes involving "research and systems-building in areas such as document retrieval, text classification, fraud detection, recommendation systems, personalized search, social network analysis, planning, diagnostics" to name a few, providing the success behind companies such as Amazon, Google, and so on. Jordan (2019) argues that referring to this work as "human-imitative AI" or AI, "prevents a clear understanding of the range and commercial issues at play", pointing to the notion of "Intelligence Augmentation" (IA) where, using the example of search engines "computation and data are used to create services that augment human intelligence and creativity." Jordan (2019) also points to "Intelligent Infrastructure" (II) described as "a web of computation, data, and physical entities that makes human environments more supportive, interesting, and safe" that may involve the Internet of Things (IoT). It is worth noting that, drawing on the work of Joshi (2019), Yigitcanlar et al. (2020) describe several levels of AI, one of which is "artificial narrow intelligence" that "represents all of the existing AI today" as distinct from yet to be achieved "artificial general intelligence" where it is claimed "AI agents can learn, perceive, understand, and function completely like a human being"; and "artificial superintelligence" where "AI replicates the multifaceted intelligence of human beings" and exceeds such capabilities.

Li, Batty, and Goodchild (2020) claim that "information is indispensable to the successful management and operation of cities" enabled through "the widespread adoption of computers, communication networks, and sensors" and the abundance of "computational power." To this end, Li et al. (2020) explore the role of "geographic information science and systems (GIS)" in the "effort to 'smarten' the city" focusing on the five dimensions of information and communication technologies (ICT), real-time geospatial data in cities, real-time urban simulation, real-time visualization of city-data, and real-time GIS processing. In this way, "hidden patterns" may be revealed through visualizations while enabling inspection of various elements. Al Ridhawi, Otoum, Aloqaily, Jararweh, and Baker (2020) propose a framework for "continuous, reliable, secure, and sustainable diversified smart city services" using "a Mobile Edge Computing (MEC) solution" for "next generation networks" enabling "node collaboration among IoT devices and the fog layer" as well as "the cloud layer."

Emotion/Affect

Scherer (2005) speaks in terms of a semantic space for emotions, where anger is located in the upper right active/aroused and obstructive zone and comfort as part of contentment is located in the lower left passive/calm and positive zone. Guthier, Abaalkhail, Alharthi, and El Saddik (2015) articulate the research field of "affect-aware cities" as a space for emotion research in smart city initiatives. Guthier et al. (2015) note that, "as the city evolves" opportunities arise, such that, "we are no longer planning cities" and instead, "we are reinventing, reusing, and living within them." Indeed, according to Guthier et al. (2015), "applications in smart city initiatives" enable an "affect-aware city that is able to understand, interpret and adapt" to the emotions of people. Moreira de Oliveira (2015) proposes a research plan to explore emotion-aware cities in relation to ambient geographic information (AGI). Emotion/affect is said to be important in the context of human-computer interaction, as "a critical element of design for human experience" (Hanington,

2017), extended in this work to include smart environments and regions. If as Brighenti (2017) claims, the visible involves the sensory, then this gives rise to opportunities for exploring emotion and affect in smart cities. Nielek, Ciastek, and Kopeć (2017) claim that "emotions are indispensably connected with peoples' lives" and a more inclusive platform is proposed and tested to explore the emotions of people in urban spaces since the "understanding of interaction between people and urban spaces" and how people feel about such spaces is said to be "crucial for inclusive decision making." Gross, Levenson and Mendes (2020) indicate that affective science "encompasses a broad range of affective processes" including "emotions, moods, preferences, attitudes, value-based decisions, and stress responses." Affect, according to Gross et al. (2020), "determines what we attend to and remember" and much more. And there is what the Emotional AI Lab (2020) with Andrew McStay, team, and collaborators refer to as emotional AI, said to be "technologies that use affective computing and artificial intelligence techniques to sense, learn about and interact with human emotional life." However, emotional AI is said to be "a weak form of AI" and this is because the technologies involved "aim to read and react to emotions through text, voice, computer vision and biometric sensing, but they do not have sentience or emotional states themselves" giving rise to many questions posed by the Emotional AI Lab.

Imagination

It could be argued that the fictional work by Calvino (1972; translation to English 1974) on invisible cities serves, through the imagination, to make invisible cities visible, wherein Marco Polo "describes the cities visited on his expeditions" to Kublai Khan who "does continue listening to the young Venetian with greater attention and curiosity than he shows any other messenger or explorer of his."

Infrastructure

Dourish and Bell (2007) highlight how "infrastructure, rather than being hidden from view, becomes visible through our increasing dependence on it for the practices of everyday life." Harrison and Donnelly (2011) claim that the "Smart City approach" contributes to "making the invisible visible" and enables "access to real-time information at the level of individual citizen's choices and actions" pointing to "the emerging roles of information and technology."

Silva, Vaz de Melo, Almeida, Salles, and Loureiro (2012) describe the use of participatory sensor networks (PSN) as "a very fundamental tool to study social behavior at a large scale." Based on the example of Foursquare as a location sharing service, Silva et al. (2012) "propose a technique called city image" as "a way of summarizing the city dynamics based on transition graphs, which map the movements of individuals in a PSN." Branchi, Fernández-Valdivielso, and Matias (2014) claim that, "urban modeling allows the presentation of different invisible aspects of the city which influence citizens' behavior" as well as "the shape of the city's physical form." Branchi et al. (2014) propose an analysis matrix for smart cities to assess the usefulness and consequences of different technologies.

Information

Beukes (2015) provides a description of "the social and technical complexities in achieving a single, globally accessible platform for 'slum' data." Schmitt (2015) claims that, "information architecture serves as a metaphor to structure the vast amounts of data produced in modern society" in the context of "the virtual." As such, "information architecture stands for making the invisible visible in the form of digital information extracted from and applied to physical architecture, to better understand and design" the latter. Schmitt (2015) uses the example of projections on the facades of buildings that "can become large information displays" along with "infrared cameras to detect people and heat emitting objects, as they were moving in front of the building, resulting in dynamic changes in the projections" thus "making the invisible – in this case heat sources – visible." Such experimentation is said to provide "experience in visualizing information in previously unthinkable ways." Schmitt (2017) describes the notion of visible and invisible flows of information in urban areas and regions in terms of "information cities." Sheppard (2018) argues for the importance of building literacy around issues such as climate change as well as the use of landscape architectural techniques.

Simulations

Schmitt (2015) describes the relevance of city simulations in making the invisible visible in that, "we know results of processes both from experience and observations, but we cannot directly visualize it – this is where simulation helps." For Schmitt (2015), simulation refers to "the imitation of the operation of a real-world process or system over time" enabling the potential to "test and visualise future scenarios." Morello and Piga (2016) use experiential simulation in architecture and urban space while Dembski et al. (2020) describe the use of virtual reality (VR) in the form of a 3D model as an "urban digital twin" that would "support urban planners, urban designers, and the general public" in order "to visualise the complex 'invisible', such as urban data and simulations, to support the participation of citizens or the collaboration of experts."

Electricity

Schmitt (2015) claims that, "increasingly, the building electricity infrastructure becomes invisible by design, to reduce the danger of accidentally being exposed to high voltage" such that the "only visible elements left are switches and power plugs." McCullough (2020) points to the invisibility of electricity and draws parallels with the invisible embedding of pervasive technologies providing an opportunity to "rethink many values, not only of technology but also of use."

Discourse

Kwon (2015) argues for the use of critical visualization in making the invisible visible in landscape and urban planning, as in, "Who or what are the invisibles in the dominant discourse of urban planning?" and "What visualization technologies or practices can be used to 'make visible' the interests of those who are missing or ignored in the planning process?" Writing about Hudson Yards, a smart city development in New York City, Mattern (2016) refers to *technoscientific urbanism* "that is especially visible in the discourse around 'smart cities' which regards the intelligence generated from spatial sensing and data analysis as a 'fix' for perennial urban problems." Mattern (2016) also points to the limitations of the instrumented city noting that, "if we were to measure the behavior of these citizen-sensors as an index of their engagement with the city, we'd find that their actions are limited, by the instruments that render those actions visible and worth accounting for." Caprotti and Cowley (2019) employ the "analytical approach" of discursive logics with the "fundamental aim" to "demonstrate the value of a cultural economy lens in making context-specific logics visible, with the UK used as an example." Logics are described as "a narrative device" in "textual, spoken, and image-based" form. Through "making these logics visible" the work of Caprotti and Cowley (2019) "opens up a new critical space for debate about alternative future pathways that smart cities might take." Burns and Andrucki (2020) question "who cares?" in smart cities, claiming that labor is "invisibilized in current research and public discourse" giving rise to the notion of "digital care worker" referring to "voluntary and low-paid data producers and analyzers" using the example of "hackathons" with the expectation that people will contribute "out of love for their cities and communities."

Art and Visualizations

Jeremijenko (2016) makes visible the invisible through art installations and projects such as Amphibious Architecture, combining artistic and engineering skills, enabling "audiences to visualize river health and water quality in real time with a tangible, visual representation of what is happening in their river." For example, Amphibious Architecture offers "a poetic and dynamic light array over 60 metres long" where information is provided "about river health through dissolved oxygen levels below the surface." Chevalier (2018) employs digital art in making "visible the invisible" and as such, "reveals the essence of things in an ever-changing world." Using "light, movement, energy," Chevalier (2018) articulates the "new poetics of matter" and "notions of flow and networks" in employing "the virtual the elusive" and in this way he "translates the new forms of contemporary life and cities today" providing a glimpse of "incessant renewal, speed, transformation" while "abandoning all chronological and linear memory" revealing the urban "between destruction and construction, growth and mutation." Sardo, Larbey, and Laggan (2019) advance the use of art for engaging people in healthy urban development. Wigley and Rose (2020) explore several ways in which "the real smart city" is said to "make itself visible" through visualizations such as urban dashboards, three-dimensional interactive maps, and virtual reality.

Technology

Sassen (2012) speaks in terms of urbanizing technologies, identifying the need for smart city technology to be able to "work within a particular urban context." For Sassen (2013), urbanizing is important given the deurbanization forces of inequality, the privatizing of public spaces, and systems of surveillance. McKenna (2017) extends the notion of urbanizing technologies to the urbanizing of ambient and more aware technologies. Caprotti (2017) argues that the smart city is characterized by invisibility because "projects are often computer-based and have little or no actual visible, tangible physical presence" and that people are embedded as well, as in, "were part of it" in combination with smartphones and other technology-related elements. Describing key technology trends for 2020, Gartner (Panetta, 2019) identifies human augmentation as "the use of technology to enhance a person's cognitive and physical experiences." Orasch (2019) describes the use of mapping to reveal the types of smart city technologies that "are invisible to the human eye" such as traffic noise monitors, sensors, smart trash bins, and smart benches. Sobnath and Rehman (2019) address the many ways in which the visually impaired could experience urban life through interacting with emerging technologies and applications. Shadowen, Lodato, and Loi (2020) describe an example based on the Neighborhood Networks project where participatory design (PD) was used as an approach "for residents to envision new futures for their neighborhood through prototyping, sketching, and modeling technologies" such that "design activities created an avenue for a community that felt largely invisible and marginalized in local politics and programming to become *visible* contributors to their city." Leorke (2020) advances "the conceptual categories of re-appropriating, reconfiguring, and augmenting" as mechanisms for making the smart city more visible through play. Borning, Friedman, and Logler (2020) outline and categorize the "largely invisible" impacts related to the "extracting, processing, maintaining, and ultimately disposing of the materials used to support information technology" not to mention that associated with "producing the energy" for "devices in operation, as well as in their production and disposal." Approaches provided for making the invisibilities of IT (Information Technology) visible are perhaps best understood in terms of the categories provided by Borning et al. (2020) such as mental models, metaphors, visions, and so on. Kummitha (2020) advances a sociomateriality approach to understanding the "human-technology connect in smart cities" claiming that "the greater distance between technology development and its adoption in smart cities, the less effective the outcomes" giving rise to the need to "promote the enhancement of citizens' skills and competence, placing them in a better position to develop and appropriate locally developed technologies." Ahad, Paiva, Tripathi, and Feroz (2020) provide a review of a wide range of enabling technologies for smart cities in terms of "technical, social, and economical aspects" while contributing "a comprehensive roadmap for achieving a smart city ecosystem" with an emphasis on sustainability and best practices for multiple stakeholders. Cottrill,

Jacobs, Markovic, and Edwards (2020) refer to the Internet of Things (IoT) deployments in smart cities as being "hidden in plain sight" due to "inadequate attention given to the communication of practices regarding the collection, use and sharing of data" giving rise to the challenge of "how to provide this information in a manner that is informative, visible, and accessible." In response, Cottrill et al. (2020) propose "designing for privacy and trust" in IoT deployments as an approach.

Human Geography

From a human geography perspective, Bian (2020) provides an introduction to a series of articles on smart spaces and places, noting that, "novel data and their associated sensors capture social processes and experiences that would have not been known to us otherwise."

Figure 1 provides an overview of approaches for making the invisible visible in smart cities emerging from the research literature and encompasses art, computational intelligence (CI) / communication networks (CN), geospatial analysis, discourse, human geography, electricity, infrastructure, participatory design, emotion/affect, imagination, information, simulation, and technology.

Figure 1. Overview of approaches for making the invisible visible in smart cities

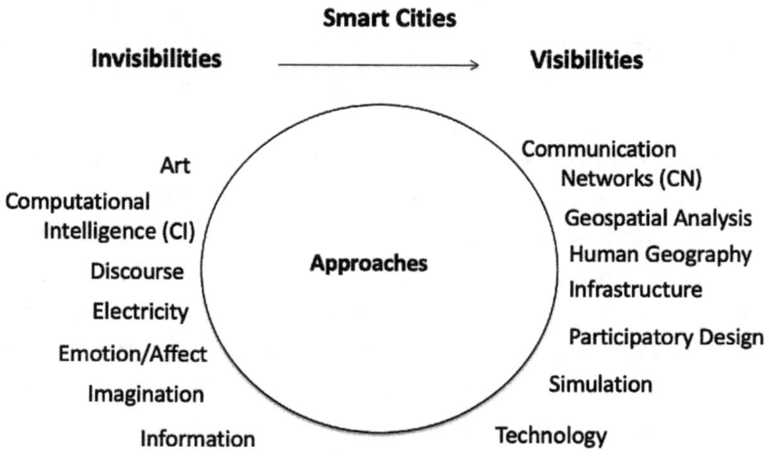

3.2 Key Approaches to Making the Invisible Visible in Smart Cities

Not to be omitted from any discussion of making the invisible visible is the fictional work of the imagination by Calvino (1972 Italian edition, 1974 English edition) on invisible cities where the narrator of the work makes cities come alive, focusing on a range of themes from trading cities to continuous cities to hidden cities.

In summary, Table 1 provides an overview of key approaches identified in the research and practice literature for making the invisible visible in smart cities, organized by year from 2007 to 2020.

Looking at Table 1, approaches to making the invisible visible in cities, and more particularly in smart cities, include but are not limited to: everyday interactions with urban infrastructures (Dourish and Bell, 2007); real-time access to information enabled by technologies (Harrison and Donnelly, 2011); participatory sensor networks (PSNs) and visualizations of the city through mapping movements and dynamics (Silva et al., 2012); community-driven data collection process for shack/slum dwellers international (SDI) (Beukes, 2015); landscape architectural techniques and literacy (Sheppard, 2018); urban infrastructure design (McCullough, 2020; Schmitt, 2015); information architecture, information visualization, and simulation (Schmitt, 2015); urban discourse and instrumentation (Mattern, 2016); art and engineering for information sharing, analysis, and visualization (Jeremijenko, 2016); computer-based projects involving people (Caprotti, 2017); urban information flows (Schmitt, 2017); Bristol's Data Dome and Glasgow's Operations Centre (GOC) as examples of urban spaces rendering smart cities visible

(Caprotti, 2018); dynamic, digital urban art (Chevalier, 2018); discursive analytics (Caprotti and Cowley, 2019); the use of mapping technologies to reveal smart city technologies (Orasch, 2019); applications for the visually impaired (Sobnath & Rehman, 2019); use of art for engaging people in healthy urban development (Sardo et al., 2019); simulation of secure and reliable communication for next generation networks using a Mobile Edge Computing framework (Al Ridhawi et al., 2020); multidimensional review of enabling technologies for sustainable smart cities (Ahad et al., 2020); novel data revealing social processes and experiences from a geographic perspective (Bian, 2020); understanding the invisible impacts of information technologies (Borning et al., 2020); discourse analysis to render visible digital caring laborers (Burns and Andrucki, 2020); designing IoT deployments for privacy and trust (Cottrill

Table 1. Overview of key approaches to making the invisible visible in smart cities

Author(s)	Year	Approaches to Making the Invisible Visible in Smart Cities
Dourish & Bell	2007	Urban infrastructures and use in everyday life
Harrison & Donnelly	2011	Information and technology affording access in real time
Silva et al.	2012	PSNs & Visualization – city image – mapping movement & dynamics
Branchi et al.	2014	Analysis matrix to assess usefulness and consequences of technologies
Beukes	2015	Community-driven data collection process for SDI
Kwon	2015	Critical visualization
Schmitt	2015	- Information architecture for digital information about the physical - Information visualization projected onto urban structures / dynamic - Simulation for visualization of scenarios - Invisible electricity infrastructures by design
Mattern	2016	Discourse – Techoscientific urbanism; Instrumented city & behavior
Jeremijenko	2016	- Art + engineering for information sharing and visualization - Urban data gathering and analysis
Caprotti	2017	Computer-based projects with people embedded as part of it
Schmitt	2017	Information flows as Information cities
Caprotti	2018	Spaces and places rendering visible the invisible (Data Dome and GOC)
Chevalier	2018	Digital art revealing the city in motion
Sheppard	2018	Landscape architectural techniques; building literacy
Caprotti & Cowley	2019	Analytical using Discursive logics
Orasch	2019	Mapping smart city technologies
Panetta	2019	Human augmentation
Sardo et al.	2019	Art for engagement in healthy urban development
Sobnath & Rehman	2019	Visually impaired
Ahad et al.	2020	Multidimensional review: enabling technologies for sustainable smart cities
Al Ridhawi et al.	2020	Simulations for a MEC solution for secure & reliable smart cities
Bian	2020	Geographic: Novel data – sensors – social processes – experiences
Borning et al.	2020	Mental models, visions, norms, etc.: impacts of information technologies
Burns & Andrucki	2020	Discourse analysis about invisibilized digital care workers
Cottrill et al.	2020	Designing for privacy and trust in IoT deployments
Kummitha	2020	Sociomateriality, technology-human connect
Leorke	2020	Play
Li et al.	2020	CICN and Real-time GIS and geospatial analysis
McCullough	2020	Invisible electricity infrastructures: opportunities to rethink tech value-use
Shadowen et al.	2020	Becoming visible through participation
Wigley & Rose	2020	Visualizations for making the "real smart city" visible

et al. 2020); play (Leorke, 2020); sociomateriality approach in support of a people-technology connect (Kummitha, 2020); computational intelligence and communication networks (CICN) in support of a real-time paradigm for GIS-enabled smart cities and geospatial analysis (Li et al., 2020); scrutiny and rethinking of invisible electricity infrastructures (McCullough, 2020); visibility through participation and the use of participatory design to enable the marginalized to become visible in policymaking and planning processes (Shadowen et al., 2020); and visualizations for making the 'real smart city' visible (Wigley and Rose, 2020).

This review of the research and practice literature provides a range of perspectives on making the invisible visible, enabling emerging understandings of the multiple components of smart cities pertaining to data; information; infrastructures; art; aware and interactive technologies such as sensors; discourse; social processes; and human behavior, to name a few. Such approaches may be key to consider for those involved with smart city interventions, initiatives, and projects going forward.

Across this vast array of approaches, what emerges more than once is art, information, technology, data, mapping, visualization, and infrastructures (visible and invisible). Also of note is human augmentation and sensors for sensing. These terms will inform and guide the selection of elements or proxies used in exploring relationships for making the invisible visible in smart cities in this chapter.

3.3 A Conceptual Framework for Making the Invisible Visible in Smart Cities

Based on the review of the research literature presented in this chapter, an emerging conceptual framework for making the invisible visible, in the context of smart cities and regions, is advanced as depicted in Figure 2.

Heightening urban sensibilities and *emotion/affect* in the form of human tension and *anger* as invisibilities are made visible using *urbanizing* as evolving approaches such as augmentations, urban displays, and visualizations, giving rise to the potential for greater awareness for collaborating involving digital and virtual elements in research and practice opportunities.

The research question in Section 1 is reformulated here as a proposition for exploration in this chapter, as follows:

Figure 2. Making the invisible visible: An emerging conceptual framework for smart cities and regions

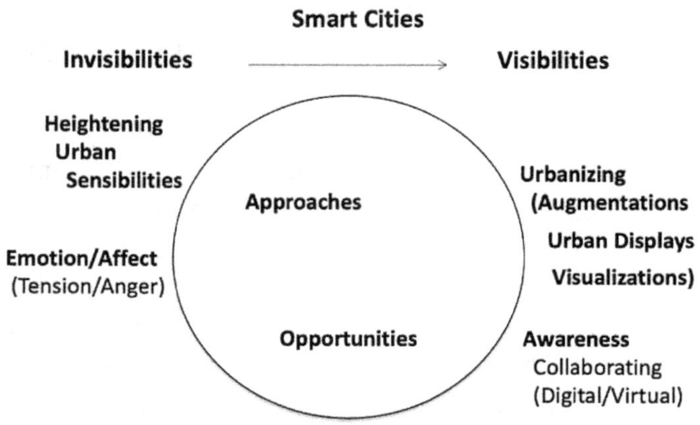

P: Making the invisible visible in smart cities and regions provides greater understandings of the potentials enabled by heightening urban sensibilities for urbanizing, as in, adapting for urban uses, and emotion/affect such as feelings of tension or anger in gaining greater awareness capabilities for collaborating and interacting with each other and with aware technologies (digital/virtual) as opportunities important in evolving research and practice.

4. AN EXPLORATION OF MAKING THE INVISIBLE VISIBLE IN SMART CITIES

This chapter undertakes an exploration of the experiences of people in thinking about and assessing their city as smart in terms of how the invisible is being made visible using an exploratory case study approach combined with an explanatory correlational design.

4.1 The Design for Exploring Making the Invisible Visible in Smart Cities

Through an exploratory case study approach where the experience of people from diverse sectors in multiple small to medium to large-sized cities forms the case for this study, people were invited to participate through a website developed for the purposes of the study. The case study approach is combined with an explanatory correlational design to explore the relationships between urban experiences based on *heightening urban sensibilities* and *urbanizing* (e.g., adapting for urban uses). The study was described through a website and during sign up demographic data were collected including elements such as age range, location (city), and people could self identify in one or more categories (e.g., community member, city official, researcher, educator, student, business, etc.). Upon sign up, people were invited to complete a survey about their experience of their city as a smart city. An invitation was also provided to participate in an in-depth interview about smart cities, online or in person, depending upon location. The study attracted participation from across Canada and extended to other cities in the United States, Ireland, Greece, Israel, and so on. The survey instrument, containing a mix of closed and open-ended questions, and the interview protocol were pretested prior to use in the study. To enable responses from a broad spectrum of people, heterogeneity sampling, a type of purposive sampling was used whereby an invitation to participate in the study was posted to webspaces with the potential to attract a wide response. Data analysis included the use of content analysis for qualitative data and descriptive statistics for quantitative data. The Real Statistics Software Resource Pack Add In (Zaiontz, 2019) for Microsoft Excel was used, consisting as it does of a wide range of tools for statistical analysis, such as the correlation feature for calculation of the Spearman correlation for the ordinal data, generated by assessments from survey respondents. Concurrent with this study, data were also collected through group and individual discussions with a wide range of people across many sectors in small to medium to large-sized cities in Canada (e.g., greater Victoria, Vancouver, and Toronto). Overall, this chapter is based on an analysis of data for n=78 including 41% females and 59% males for people ranging in age from their 20s to their 70s.

4.2 Exploring Practices for Making the Invisible Visible in Smart Cities

A range of participants described ways for making the invisible visible in their city. For example, the notion of the invisible dimension of human connection in urban spaces was described in relation to a fountain that was said to be a "touchstone" for the community. An educator spoke of the fountain as an urban element that "brought people out" and "made people talk" with each other, indicating also that, "it slowed people down" and "it was fun." Additionally, the notion of an invisible element in urban spaces as that of, how one feels in the city, was described by a community leader in relation to the placement of urban elements such as benches. For example, a community member and leader commented that, "the placement and proximity of those benches created that opportunity" for "conversation with the person that was sitting across from me." The notion of smartness was described by one individual as, "nothing visible" however, it was observed that urban displays showing real time, interactive counts for cyclists came to mind as an example of smartness. Interest was expressed by another community leader in having count information for "public spaces, like plazas" to learn more about "some of that anonymous kind of count" data to reveal "what the daily average" of usage might be.

An educator described smartness as "smart meters for footfall, air and water quality" while noting that, "urban displays become more and more present" identifying the concern that "content is merely advertising." However, the educator observed that, "the number of places available in each city centre car park is displayed at all major junctions" while "real time bus data is available on displays at almost each stop" and "mobile phones are everywhere" so that "one can pay for parking via mobile phone" and "check the shops' opening hours, the cinema shows, etc." Another educator spoke of "the visual sense of the city" in relation to what cities do for the creation of this dimension. A community leader described a design thinking approach to collaboratively figuring out what might be possible in terms of adaptive reuse or "upcycling" of one or more fragments of an urban infrastructural element with iconic and historical value. City information technology (IT) staff pointed to the shared use of drones for emergency services across municipalities, in making visible, aerial views and mappings of the city and for, "search and rescue and safety operations." And an educator noted that, "drone footage is interesting for providing a bird's eye view of major events."

Quantitatively, when asked to assess whether city-focused social media and other aware technologies give rise to many possibilities, people responded to several items for this survey question using a 7-point Likert-type scale (e.g., 1 = Not at all, 2 = Not sure, 3 = Maybe, 4 = Neutral, 5 = Sort of, 6 = Sure, 7 = Absolutely). For example, Table 2 shows assessments for the items *heightening urban sensibilities* and *urbanizing (e.g., adapting for urban uses)* in smart cities. For heightening urban sensibilities, 33% of respondents selected position 5 on the scale, 17% position 6, and 50% selected 7 on the

Table 2. Assessments for heightening urban sensibilities and urbanizing in smart cities

Items	Assessments	Correlation
Heightening urban sensibilities	33% (5); 17% (6); 50% (7)	.43
Urbanizing (e.g., adapting for urban uses)	50% (5); 17% (6); 33% (7)	

Figure 3. Scatterplot for heightening urban sensibilities and urbanizing in smart cities

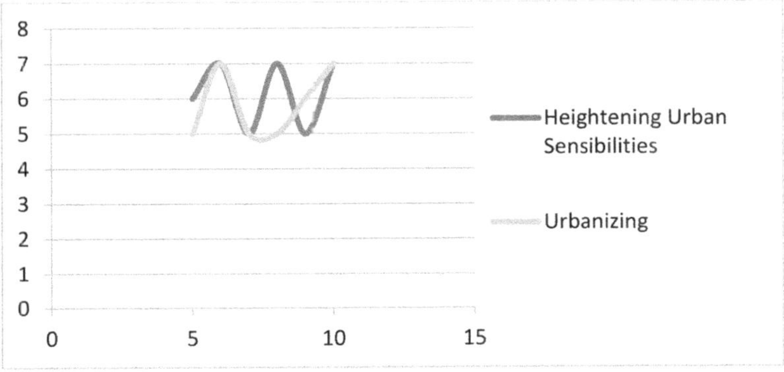

scale. For urbanizing, 50% of respondents selected position 5 on the scale, 17% of respondents selected 6 on the scale, and 33% selected 7 on the scale.

Looking at the scatterplot for this relationship, as heightening urban sensibilities increases or decreases, so too does urbanizing, for the most part, as depicted in Fig. 3.

This assessment enabled the correlating of *heightening urban sensibilities* with *urbanizing (adapting for urban uses)* where, using the Real Statistics Software (Zaiontz, 2019) Resource Pack add-in for Microsoft Excel, which contains a variety of statistical functions and data analysis tools, the correlation feature was used to calculate the Spearman correlation resulting in a correlation of .43.

Anger as emotion/affect was then correlated with *the ambient* using the proxy of *awareness*, as indicated in Table 3. Survey respondents were asked to assess an item from Anderson's Body Insight Scale (BIS): Regarding your experience in your city, "I can feel my body tighten up when I am angry"* (On a scale of 1 to 5 where 1 = strongly disagree and 5 = strongly agree). As indicated in Table 3, 17% responded at the neutral position of 3 on the scale, 50% at position 4, and 33% at position 5. When asked to assess to what extent elements such as awareness contribute to the making of smart cities,

Table 3. Assessments for emotion/affect (anger) and the Ambient (awareness) in smart cities

Items	Assessments	Correlation (Spearman)
Emotion/affect (anger)	17%; (3); 50% (4); 33% (5)	.81
The Ambient (awareness)	25% (5); 25% (6); 50% (7)	

Figure 4. Scatterplot emotion/affect (anger) and the Ambient (awareness) in smart cities

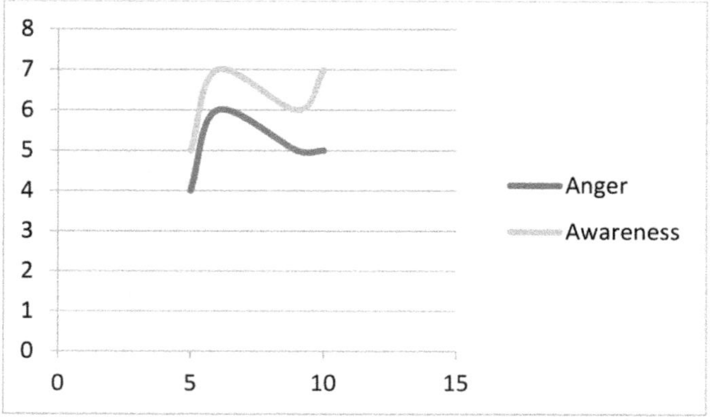

25% responded at position 5 on the 7-point Likert-type scale, 25% at position 6, and 50% at the upper end of the scale at position 7.

Anger explored as emotion/affect was then correlated with *the ambient* using the proxy of *awareness*, resulting in a Spearman correlation of .81. Looking at the scatterplot for this relationship in Fig. 4, as awareness increases or decreases, so too does a of sense of anger, for the most part.

Findings of a qualitative nature that emerged to inform the quantitative findings pertaining to emotion/affect (e.g., tension such as anger) in relation to awareness in the city include a range of comments.

Anger. Probing into neutral responses to assessments of anger in the city, one individual commented that in a smaller city context "you might get angry if you can't find a parking place but that would be as serious as it got" while on a much larger scale, "conversely though you would be so used to being angry or tough that your body would not tighten up because it would be normal." In a larger city, this individual commented that, "you learn an entirely different way of living in a city like that" highlighting "this difference in how you carry yourself."

Awareness. Referring to the physical and the digital in terms of finding a restaurant, one community member commented that, "I guess it's kind of a tradeoff of awareness between what's around or what I can perceive without the technology to what is perceived through the technology." An educator referred to the "concurrent awareness" occurring across "totally different places in the globe" as in "this seamless interrelationship of your local and your global" where "people are sharing some event that is going on around them with people that they know" elsewhere in the world "and you just take it for granted" as being "completely normal." A public administrator spoke of the importance of "being aware" of being a "citizen of the city." A community leader indicated that, "the biggest benefit" of "social media and other aware technologies" is the ability "to assist in citizen/visitor education and awareness" while acknowledging that, "unfortunately, most cities are not utilizing them in this manner."

4.3 Discussion of Potentials for Making the Invisible Visible in Smart Cities

Both quantitative and qualitative approaches emerge in this exploration of practices for making the invisible visible in smart cities and regions. For example, approaches that emerged in this exploration include that of count methodologies and engaging people with being attentive to the importance and value of fun, play, art, learning, design thinking, and the like. Quantitatively, correlation is used as a way or an approach to learn more about the nature of the relationship between urban activities such as *heightening urban sensibilities* and *urbanizing*, as in, *adapting for urban uses*. Although the correlation of .43 achieved in this work, based on assessments of *heightening urban sensibilities* and *urbanizing* in smart cities, is said to be "useful for limited prediction" according to Creswell (2018), correlations in the range of .35 to .65 provide "values used to identify variable membership in the statistical procedure of factor analysis" for the "intercorrelation of variables" and this could be helpful for future smart city initiatives, projects, and interventions. Further, approaches identified in the qualitative data have the potential to enhance, enrich, and activate the people dimension of smart cities and regions. Regarding emotion/affect, as indicated by an educator, city size seems to matter where, in a small city the awareness of anger was associated with not being able to find a parking space, whereas in a larger city the human body is said to be in a constant state of awareness and on the alert for any number of things that could give rise to tension and anger. Regarding the *anger* and *awareness* correlation of .81, according to Creswell (2018), a correlation between .66 and .85 indicates a "very good" relationship and "good prediction can result from one variable to the other."

In practical terms, possibilities for the urbanizing of technologies, as in, adapting for urban uses, across municipal boundaries emerged in relation to the use of drones in the provision of mapping, emergency, and other services. Where urban displays and digital signage have tended to focus on advertising, findings in this work identify the opportunity for utilizing these urban elements for much more than "merely advertising" suggestive of the need for much more interactive urban displays with meaningful content focused on informing and learning. Additionally, these and other opportunities emerge for urban elements to include visualizations that contribute to the "visual sense of the city." The pervasive presence of smart phones and their use enables the dynamic and interactive generation of current information about a wide range of urban services, events, and the like to inform planning, decision-making, and overall awareness. As such, smart cities and regions involve much broader understandings of "smartness" extending to the human elements of *heightening urban sensibilities*, *urbanizing* involving people in activities such as "upcycling" as in, adapting for urban uses, and the ongoing identification of other potential variables for inclusion and assessment. The emerging conceptual framework advanced in this chapter for making the invisible visible offers opportunities for researchers, serving as it does as a learning tool for designing urban spaces on the one hand while contributing also to practice, as an engagement and guidance tool on the other hand. For example, elements of the framework such as the digital and virtual can become experimental spaces for augmentations, urban displays, visualizations, and so on.

5. EMERGING RESEARCH AND OPPORTUNITIES

In this section, emerging and future opportunities are identified for practice and research pertaining to making the invisible visible in smart cities and regions.

5.1 Research

This chapter advances the potential for future research opportunities and emerging discourse spaces in terms of approaches for making the invisible visible in the context of smart cities and regions by inviting explorations of the viability of the framework operationalized for use in this chapter and as such, provides opportunities for further testing and validation. How cities are designed with people in mind, while involving people more meaningfully is highlighted, focusing for example on interactive urban elements such as fountains, benches, urban displays, mobile data applications (apps), and the like.

Approaches

Making as a form of urbanizing, as in, adapting for urban uses provides an approach to exploring further the viability of the framework for making the invisible visible conceptualized for use in this chapter. While the framework demonstrates potentials, it also provides avenues for testing and practice through offering and opening up new understandings for evolving approaches to smart city urban spaces and regions.

Making Invisibilities Visible

Conceptualization of spaces for real-time, dynamic, and adaptive types of approaches to making the invisible visible emerge in this chapter opening the way for new techniques and possibly novel roles and collaborations for community members and people in businesses, researchers, students, and governments. As such, areas for exploration potential include:

1. *Augmentations* supporting extended understandings of urban visualizations and the visual sense of the city through drone coverage and the like
2. *Urban Interactive Displays* that support more interactivity and meaningful content contributing to greater awareness, learning, and informing in real time about the city
3. *Urban Elements* involving understandings of the importance of design, participatory design, and purposeful engagement that extends to people

and their interactions in urban environments and regions through public fountains, benches, and the like

4. *Emotion/affect (e.g., feelings of tension in the city) Anger* explored through urban initiatives focusing on awareness in public spaces taking people and their experiences into consideration and possibly utilizing dynamic and interactive assessments to learn more about what contributes to stress in urban environments and regions

5.2 Practice

On a practical level, this chapter contributes to spaces for the rethinking of approaches in addressing opportunities associated with making the invisible visible in smart cities and to the surfacing of associated variables and indicators.

Making Invisibilities Visible

This chapter provides exposure to, and understandings of, urban approaches to making the invisible visible in smart cities and as such, points to emerging potentials for the use of urban interactive displays, urban elements such as benches, and augmentations of municipal services such as in the use of drones for mapping or in assessing emergency situations.

Variables and Indicators

This chapter surfaces the variables of *heightening urban sensibilities* and *urbanizing* in the form of *adapting for urban uses* in the context of visibilities and invisibilities and associated variables related to augmentations (e.g., making benches smart), ways of creating a visual sense of the city, and then enhancing of any number of urban elements such as urban interactive displays, and so on.

6. CONCLUSION

This chapter provides an overview from the research and practice literature of approaches for making the invisible visible in smart cities, encompassing a range of disciplinary perspectives spanning information technology to human geography to information architecture. From the case study exploration conducted in this work, urban interactive displays, the placement of urban elements such as benches, and augmentations in the form of drones emerge as key elements for making the invisible visible and for navigating new approaches and conceptualizations for smarter cities and regions. The main contributions of this chapter include a) articulation and operationalization of an emerging conceptual framework for making the invisible visible in smart cities; and b) advancing of the value of variables such as *heightening urban sensibilities, emotion/affect* (e.g., human tension in the form of *anger*), and *urbanizing* using *adapting for urban uses* as a proxy for involving people more meaningfully with smart cities and regions in making the invisible visible in fostering smarter, more responsive, cities. Such variables could also be explored further going forward for their potential in contributing to the success of initiatives, interventions, and projects for smart cities and regions.

Limitations of this chapter may pertain to the use of *adapting for urban uses* as a proxy for *urbanizing* in exploring urban relationships for making the invisible visible in smart cities and this is mitigated by examples from the research literature (Sassen, 2012) and in-depth discussions with a range of individuals regarding everyday emergent practices and opportunities and the potential to extend this study to include other variables. Further limitations related to literature review domain coverage are mitigated by the interdisciplinary potential to extend this review to other areas going forward.

This chapter continues to give form, detail, and definition to the invisibilities and visibilities constructs in relation to human sensibilities, augmentations, urban interactive displays, and the placement of urban elements.

Insights of particular interest to the reader include the following:

1. The emerging range of approaches to making the invisible visible in evolving understandings of smart cities
2. The presence of a "useful" correlation between *heightening urban sensibilities* and *urbanizing* in smart cities and the presence of a "very good" correlation between *emotion/affect* in the form of feelings of *anger* and *awareness* with good prediction potential
3. The interplay of augmentations, urban interactive displays, and urban elements in making visible the invisible

Ideas highlighted in this chapter pertain to the following:

1. The importance of people becoming more aware of the invisible and intangible aspects of the smart cities phenomena

2. The potential for people to shed light on rendering visible, invisible spaces in digital/physical urban environments
3. The potential for opportunities going forward to involve people more meaningfully in making the invisible visible

Exercises for the reader to consider, encouraging further explorations, include the following:

1. Identify one or more examples of making the invisible visible in your city
2. Identify one or more approaches for making the invisible visible in your city based on your example
3. Describe why it could be important to make the invisible visible, based on your example

If you would like to share your responses to the Exercises with the author, please complete the questions in this online space for Chapter 2 – Making the Invisible Visible

This work will be of interest to a broad audience including students, educators, researchers, artists, urban planners, urban practitioners, community members, and anyone concerned with emerging approaches and practices that take into consideration the complex urban opportunities posed in rendering the invisible visible in smart, responsive, and future cities.

REFERENCES

Ahad, M. A., Paiva, S., Tripathi, G., & Feroz, N. (2020). Enabling technologies and sustainable smart cities. *Sustainable Cities and Society*, *61*, 102301. doi:10.1016/j.scs.2020.102301

Al Ridhawi, I., Otoum, S., Aloqaily, M., Jararweh, Y., & Baker, T. (2020). Providing secure and reliable communication for next generation networks in smart cities. *Sustainable Cities and Society*, *56*, 102080. doi:10.1016/j.scs.2020.102080

Belanger, L., & Hinchliffe, E. (2020). 5 big ideas for fixing global cities' most daunting challenges. *Fortune Magazine*. Retrieved 19 February 2020 from https://fortune.com/2020/02/17/cities-challenges-solutions-housing-buses-climate-loneliness/

Beukes, A. (2015). *Making the invisible visible: Generating data on 'slums' at local, city, and global scales*. Working Paper. London: International Institute for Environment and Development (IIED).

Bian, L. (2020). Introduction: Smart spaces and places. *Annals of the Association of American Geographers, 110*(2), 335–338. Advance online publication. doi:10.1080/24694452.2019.1702810

Borning, A., Friedman, B., & Logler, N. (2020). The 'invisible' materiality of information technology. *Communications of the ACM, 63*(6), 57–64. doi:10.1145/3360647

Branchi, P. E., Fernández-Valdivielso, C., & Matias, I. R. (2014). Analysis matrix for smart cities. *Future Internet, 6*(1), 61–71. doi:10.3390/fi6010061

Brighenti, A. M. (2017). The visible: Element of the social. *Frontiers in Sociology, 2*(17), 1–11. doi:10.3389/fsoc.2017.00017

Burns, R., & Andrucki, M. (2021). Smart cities: Who cares? *Environment and Planning A. Economy and Space, 53*(1), 12–30. doi:10.1177/0308518X20941516

Calvino, I. (1974). Invisible cities. New York: Harcourt. (Original publication 1972)

Caprotti, F. (2017). Research in the invisible city: Challenges for 'knowing' in the smart city. *UGEC Viewpoints: A Blog on Urbanization and Global Environmental Change*. Retrieved 17 June 2018 from https://ugecviewpoints.wordpress.com/2017/04/20/research-in-the-invisible-city-challenges-for-knowing-the-smart-city/

Caprotti, F. (2018). Spaces of visibility in the smart city: Flagship urban spaces and the smart urban imaginary. *Urban Studies (Edinburgh, Scotland)*, 1–15. doi:10.1177/0042098018798597

Caprotti, F., & Cowley, R. (2019). Varieties of smart urbanism in the UK: Discursive logics, the state and local urban context. *Transactions of the Institute of British Geographers, 44*(3), 587–601. doi:10.1111/tran.12284

Chevalier, M. (2018). *Meta-Cities 7, Ubiquity 1 & 2*. Retrieved 16 February 2020 from https://www.mayorgallery.com/exhibitions/526/overview/

Cottrill, C. D., Jacobs, N., Markovic, M., & Edwards, P. (2020). Sensing the city: Designing for privacy and trust in the Internet of Things. *Sustainable Cities and Society, 63*, 102453. doi:10.1016/j.scs.2020.102453

Creswell, J. W. (2018). *Educational Research: Planning, Conducting, and Evaluating Quantitative and Qualitative Research* (6th ed.). Pearson.

Dembski, F., Wössner, U., Letzgus, M., Ruddat, M., & Yamu, C. (2020). Urban digital twins for smart cities and citizens: The case study of Herrenberg, Germany. *Sustainability*, *12*(6), 2307. doi:10.3390u12062307

Dourish, P., & Bell, G. (2007). The infrastructure of experience and the experience of infrastructure: Meaning and structure in everyday encounters with space. *Environment and Planning. B, Planning & Design*, *34*(3), 414–443. doi:10.1068/b32035t

Emotional AI Lab. (2020). What is emotional AI? *Emotional AI Lab: Ethics, Society, Culture*. Retrieved 1 June 2020 from https://emotionalai.org/so-what-is-emotional-ai/

Feldman Barrett, L., Mesquita, B., Ochsner, K. N., & Gross, J. J. (2007). The experience of emotion. *Annual Review of Psychology*, *58*(1), 373–407. doi:10.1146/annurev.psych.58.110405.085709 PMID:17002554

Gross, J., Levenson, R. W., & Mendes, W. B. (2020). Affective Science. *Affec Sci*, *1*(1), 1–3. doi:10.100742761-020-00008-8

Guthier, B., Abaalkhail, R., Alharthi, R., & El Saddik, A. (2015). The affect-aware city. *Proceedings of the International Conference on Computing, Networking and Communications*, 630-636.

Hanington, B. (2017). Design and emotional experience: Introduction. In M. Jeon (Ed.), *Emotions and affect in human factors and human-computer interaction* (pp. 165–183). Elsevier. doi:10.1016/B978-0-12-801851-4.00006-9

Harrison, C., & Donnelly, I. A. (2011). A theory of smart cities. In *Proceedings of the 55th Annual Meeting of the International Society for the Systems Sciences*. University of Hull Business School.

Hill, D. (2008). The Adaptive City. *City of Sound*. Retrieved 17 February 2020 from https://www.cityofsound.com/blog/2008/09/the-adaptive-ci.html

Jeremijenko, N. (2016). *Natalie Jeremijenko*. GASP. Retrieved 6 February 2020 from http://gasp.org.au/2015/12/natalie-jeremijenko/

Joshi, N. (2019). 7 types of artificial intelligence. *Forbes*. Retrieved 18 April 2021 from https://www.forbes.com/sites/cognitiveworld/2019/06/19/7-types-of-artificial-intelligence/?sh=18573180233e

Kummitha, R. K. R. (2020). Why distance matters: The relatedness between technology development and its appropriation in smart cities. *Technological Forecasting and Social Change*, *157*, 120087. doi:10.1016/j.techfore.2020.120087

Kwon, M.-P. (2015). Critical visualization in landscape and urban planning: Making the invisible visible. *Landscape and Urban Planning*, *143*, 243–244. doi:10.1016/j.landurbplan.2015.07.011

Leorke, D. (2020). Reappropriating, reconfiguring and augmenting the smart city through play. In A. Nijholt (Ed.), *Making Smart Cities More Playable: Exploring Playable Cities*. Springer. doi:10.1007/978-981-13-9765-3_3

Li, W., Batty, M., & Goodchild, M. F. (2020). Real-time GIS for smart cities. *International Journal of Geographical Information Systems*, *34*(2), 311–324. doi:10.1080/13658816.2019.1673397

Mattern, S. (2014). Interfacing urban intelligence. *Places Journal*. Retrieved 1 February 2020 from https://placesjournal.org/article/interfacing-urban-intelligence/

Mattern, S. (2015). History of the Urban Dashboard. *Places Journal*. Retrieved 1 February 2020 from https://placesjournal.org/mission-control-a-history-of-the-urban-dashboard

Mattern, S. (2016). Instrumental city: The view from Hudson Yards, circa 2019 – The world's most ambitious "smart city" project is here. Should we worry that New York City is becoming an experimental lab? *Places Journal*. Retrieved 1 February 2020 from https://placesjournal.org/article/ instrumental-city-new-york-hudson-yards/

Mattern, S. (2020). Post-It Note City. *Places Journal*. Retrieved 16 February 2020 from https://placesjournal.org/article/post-it-note-city/

McCullough, M. (2020). *Downtime on the microgrid: Architecture, electricity, and smart city islands*. MIT Press. doi:10.7551/mitpress/11953.001.0001

McKenna, H. P. (2017). Urbanizing the ambient: Why people matter so much in smart cities. In S. Konomi & G. Roussos (Eds.), *Enriching urban spaces with ambient computing, the Internet of Things, and smart city design* (pp. 209–231). IGI Global. doi:10.4018/978-1-5225-0827-4.ch011

McKenna, H. P. (2019). Visibilities and invisibilities: Methodologies and theoretical spaces for ambient urbanities. In *Ambient Urbanities as the Intersection between the IoT and the IoP in Smart Cities* (pp. 192–222). IGI Global. doi:10.4018/978-1-5225-7882-6.ch007

Moore, C. (2017). The virtual Singapore project aims to digitize an entire city. *Digital Trends*. Retrieved 8 January 2018 from https://www.digitaltrends.com/home/virtual-singapore-project-mapping-out-entire-city-in-3d/

Moreira de Oliveira, T. H. (2015). The emotion-aware city: Using ambient geographic information (AGI) in order to understand emotion & stress within smart cities. In A. Comber, B. Bucher, S. Ivanovic (Eds.), *Proceedings of the 3rd AGILE Phd School*. http://ceur-ws.org

Morello, E., & Piga, B. E. A. (2016). Experiential simulation in architecture and urban space: Introduction to the special issue. *Ambiances: International Journal of Sensory Environment, Architecture and Space*, 671.

Nielek, R., Ciastek, M., & Kopeć, W. (2017). *Emotions make cities live. Towards mapping emotions of older adults on urban space. In WI-IAT*. ACM.

Orasch, A. K. (2019). Seeing the smart city: Mapping technologies in Canada. *The Conversation*. Retrieved 19 December 2019 from https://theconversation.com/seeing-the-smart-city-mapping-technologies-in-canada-112458

Panetta, K. (2019). *Gartner top 10 strategic technology trends for 2020*. Gartner Inc. Retrieved 9 February 2020 from https://www.gartner.com/smarterwithgartner/gartner-top-10-strategic-technology-trends-for-2020/

Rickert, T. (2013). *Ambient rhetoric: The attunements of rhetorical being*. University of Pittsburgh Press. doi:10.2307/j.ctt5hjqwx

Sáinz-Piña, R. M. (2011). *Smart cities: A first step towards the Internet of Things*. Fundación Telefónica.

Sardo, M., Larbey, R., & Laggan, S. (2019). Making the invisible visible: Using art to engage people in healthy urban development. Presented at SciComm South West 2019, Bristol, UK.

Sassen, S. (2012). *Urbanizing technology*. LSE Cities. Retrieved May 26, 2015, from https://lsecities.net/media/objects/articles/urbanising-technology/en-gb/

Sassen, S. (2013). Does the city have speech? *Public Culture*, *25*(2), 209–221. doi:10.1215/08992363-2020557

Schaeffler, J. (2008). *Digital signage: Software, networks, advertising, and displays: A primer for understanding the business* (1st ed.). Focal Press.

Scherer, K. R. (2005). What are emotions? And how can they be measured? *Social Sciences Information. Information Sur les Sciences Sociales*, *44*(4), 695–729. doi:10.1177/0539018405058216

Schmitt, G. (2015). *Information Cities*. https://itunes.apple.com/us/book/information-cities/id970529491?mt=11

Schmitt, G. (2017). *Future Cities*. MOOC (Massive Open Online Course). Zurich, Switzerland: ETHzurich. Retrieved 11 December 2017 from https://www.edx.org/course/future-cities-ethx-fc-01x

Servières, M., L'Her, G., & Siret, D. (2017). Mobile devices and urban ambiances: How connected wearable tools change the ways we perceive and design public spaces. In B. E. A. Piga & R. Salerno (Eds.), *Urban design and representation: A multidisciplinary and multisensory approach*. Springer. doi:10.1007/978-3-319-51804-6_16

Shadowen, N., Lodato, T., & Loi, D. (2020). Participatory governance in smart cities: Future scenarios and opportunities. *Proceedings of HCII2020*, 443-463. 10.1007/978-3-030-50344-4_32

Sheppard, S. R. L. (2018). Making climate change visible: A critical role for landscape professionals. In E. M. H. Imfield, Y. Abunnasr, & R. L. Ryan (Eds.), *Planning for Climate Change: A Reader in Green Infrastructure and Sustainable Design for Resilient Cities*. Routledge. doi:10.4324/9781351201117-40

Shilov, N., Smirnova, O., Morozova, P., Turov, N., Shchekotov, M., & Teslya, N. (2019). Digital signage personalization for smart city: Major requirements and approach. *2019 IEEE International Black Sea Conference on Communications and Networking (BlackSeaCom)*, 1-3. 10.1109/BlackSeaCom.2019.8812820

Silva, T. H., Vaz de Melo, P. O. S., Almeida, J. M., Salles, J., & Loureiro, A. A. F. (2012). Visualizing the invisible image of cities. In *International Conference on Green Computing and Communications, Conference on Internet of Things, and Conference on Cyber, Physical and Social Computing*. IEEE.

Sobnath, D., & Rehman, I. U. (2019). Smart cities could give the visually impaired a new outlook on urban life. *The Conversation*. Retrieved 8 April 2020 from https://theconversation.com/smart-cities-could-give-the-visually-impaired-a-new-outlook-on-urban-life-123478

Want, R., & Schilit, B. N. (2012). Interactive digital signage. *Computer (Long. Beach. Calif)*, *45*(5), 21–24.

Wigley, E., & Rose, G. (2020). Will the real smart city please make itself visible? In K. S. Willis & A. Aurigi (Eds.), *The Routledge companion to smart cities* (pp. 301–311). Routledge. doi:10.4324/9781315178387-21

Wißotzki, M. (2017). Digital signage and targeted advertisement based on personal preferences and digital business models. *21st Conference of Open Innovations Association (FRUCT)*, 374–381. 10.23919/FRUCT.2017.8250206

Yigitcanlar, T., Butler, L., Windle, E., Desouza, K. C., Mehmood, R., & Corchado, J. M. (2020). Can building "artificially intelligent cities" safeguard humanity from natural disasters, pandemics, and other catastrophes? An urban scholar's perspective. *Sensors (Basel)*, *20*(10), 2988. doi:10.339020102988 PMID:32466175

Zaiontz, C. (2019). *Real Statistics using Excel*. www.real-statistics.com

ADDITIONAL READING

Gordon, R. (2020). Sprayable user interfaces: CSAIL's sprayable tech system lets users create large-scale interactive surfaces with sensors and displays using airbrushed inks. *MIT News*. Retrieved 16 April 2020 from https://news.mit.edu/2020/mit-csail-sprayabletech-sprayable-user-interfaces-0408

Hamraie, A. (2018). A smart city is an accessible city. *Atlantic (Boston, Mass.)*, *6*(November). Retrieved April 7, 2020, from https://www.theatlantic.com/technology/archive/2018/11/city-apps-help-and-hinder-disability/574963/

Hunter, N. (2009). Making the invisible, visible: Exploring liminality of mind and body through glass. https://ro.ecu.edu.au/theses_hons/1322

McGregor, G. (2020). The city that sees it all. *Fortune Magazine*. Retrieved 19 February 2020 from https://fortune.com/longform/shenzens-smart-city-technology/

Raskar, R. (2016). Making the invisible visible: Matter, methods, and minds. *MIT Technology Review*. Retrieved 10 June 2020 from https://www.technologyreview.com/2016/10/14/156956/making-the-invisible-visible-matter-methods-and-minds/

Sandercock, L. (Ed.). (1998). *Making the invisible visible: A multicultural planning history*. University of California Press. doi:10.1525/9780520918573

Shneiderman, B. (2020). Data visualization's breakthrough moment in the COVID-19 crisis. *Nightingale: The Journal of the Data Visualization Society*. Retrieved 6 April 2020 from https://medium.com/nightingale/data-visualizations-breakthrough-moment-in-the-covid-19-crisis-ce46627c7db5

Tracy, K., & Covino, K. (2017). Making the invisible visible. *Hybrid Pedagogy*. Retrieved 10 June 2020 from https://www.technologyreview.com/2016/10/14/156956/making-the-invisible-visible-matter-methods-and-minds/

KEY TERMS AND DEFINITIONS

Augmentations: Augmentations refer to the extending of capabilities involving the use of technologies, as in for example, augmented reality and the visual overlay of information in everyday spaces, contributing to heightened forms of awareness.

Awareness: Awareness is conceptualized in this work as the quality of being aware as it applies to people on the one hand, to the enabling of technologies on the other, and to the enhancing of awareness through people interacting with awareness-enabled and enabling technologies.

Collaboration: Collaboration in the context of smart environments refers to more adaptive and interactive approaches enabled through aware people and awareness-enabled technologies.

Invisibilities: Invisibilities refer to urban elements that are rendered invisible or that may present as intangibles.

Smart Cities: Smart cities are urban areas and regions characterized by more aware and engaged people, interacting with each other and aided by the use of more awareness enhancing technologies for a wide range of purposes from communication to design to atmospheres to mobility to livability, to governance, to data generation and use for learning and informed decision-making.

Urbanizing: Urbanizing refers to activities associated with the adapting of urban spaces, elements, and regions for urban use.

Visibilities: Visibilities refer to physical and other visible urban elements and infrastructures.

ENDNOTE

* "Reproduction by special permission of the Publisher, Mind Garden Inc., www.mindgarden.com from the Body Insight Scale by Rosemarie Anderson, Ph.D. Copyright 2011 by Rosemarie Anderson. Further Reproduction is prohibited without the Publisher's written consent."

Section 2

Transparency, Translucence, and the Ambient

Chapter 3
Transparency and the Ambient:
Evolving Perspectives in Smart Cities

ABSTRACT

The purpose of this chapter is to explore transparency and the ambient in relation to evolving perspectives on smart cities and regions. As such, this chapter seeks to shed light on the importance of elements in urban environments informing urban visibilities and invisibilities for smart, responsive, and future cities. The research literature for transparency and the ambient is explored in this chapter in the context of visibilities and invisibilities in smart cities. The constructs of openness and attuning to urban spaces are used to explore transparency and the ambient. This chapter uses an exploratory case study approach in combination with an explanatory correlational design. This chapter makes a contribution to 1) the research literature for transparency and the ambient in smart cities, 2) the evolving of perspectives on smart cities and regions, and 3) the evolving of theory through formulation of a conceptual framework for transparency and the ambient in smart cities.

1. INTRODUCTION

This chapter seeks to increase understanding about urban visibilities and invisibilities through an exploration of transparency on the one hand and the ambient on the other hand, said to be "a continuum of awareness" (McCullough,

DOI: 10.4018/978-1-7998-3850-0.ch003

2013), in the context of evolving perspectives on smart cities and regions. According to Gil-Garcia, Zhang and Puron-Cid (2016), openness is a key dimension of smartness in government because it enables transparency that "promotes accountability and provides information for citizens about what their government is doing." Stephanidis et al. (2019) identify transparency as one of three "important features of intelligent systems" along with "understandability and accountability." Where Shadowen, Lodato, and Loi (2020) speak of transparency in terms of automated decision systems in smart cities, Mattern (2020) was critical of the proposed (and more recently withdrawn (McLeod and Zochodne, 2020)) Sidewalk Labs smart city project for the Toronto waterfront claiming it to be "designed to suggest transparency" and "allay fears with ostentatious displays of transparency." In the context of ambient intelligence (AmI) environments, Stephanidis et al. (2019) identify "personalized, adaptive, responsive, and proactive" as "fundamental characteristics" of the ambient. The investigation in this chapter is guided by the use of an exploratory case study along with an explanatory correlational design where data are collected through the use of multiple methods including survey and in-depth interviews. In parallel with this study, data are also gathered through individual and group discussions. This chapter is significant in that it explores the nature of the relationship between transparency and the ambient using the proxies of openness and attuning to urban spaces, respectively. This chapter makes a contribution in several ways, through: a) extending the research literature for invisibilities and visibilities in smart cities to include transparency and the ambient; b) the evolving of perspectives on smart cities and regions; and c) formulation of a conceptual framework for transparency and the ambient in smart cities.

2. BACKGROUND AND OVERVIEW

Sáinz-Piña (2011) identifies the importance of transparency in relation to e-government and administration services. Among motivating factors for smart governance, Scholl and AlAwadhi (2016) identify "desire for transparency" while smart city implementation challenges include seven items, one of which is "a transparent and effective decision-making process." Gil-Garcia et al. (2016) address transparency in relation to government smartness and the dimensions of openness and of integration. Lytras and Visvizi (2018) identify transparency of services as one of six factors for "long term planning" among "smart city services to deal with concerns of citizens." Stephanidis et al. (2019) caution that "transparency can lose its power if it does not have meaningful effects" and "can create privacy threats" adding that "making a system visible does not necessarily mean that it is understandable" and this "is a truly challenging task when it comes to complex systems." Streitz, Charitos, Kaptein, and Böhlen (2019) lament the lack of prominence of ambient intelligence (AmI), overtaken as it is by "the 'smart everything paradigm'" claiming that the AmI "approach has a lot to offer" in terms of "design imperatives", ethical considerations, and smart societies more generally. In response, this chapter undertakes the objective described below.

Objective: The objective of this chapter is to explore and advance new understandings of the relationship between *transparency* and *the ambient* in the context of urban visibilities and invisibilities. As such, the key research question posed is – *What is the nature of the relationship between transparency and the ambient in smart cities and regions, as dimensions of visibilities and invisibilities?*

2.1 Definitions

Definitions are presented here, based on the research literature, for key terms used in this chapter. Additional terms with definitions, as articulated by this work, are included in the *Key Terms and Definitions* section at the end of this chapter.

Ambient: McCullough (2013) describes ambient as "a continuum of awareness" and "an awareness of continuum."

Smart Cities: Townsend (2013) describes smart cities as "places where information technology is combined with infrastructure, architecture, everyday objects and even our bodies, to address social, economic, and environmental problems."

Transparency: Cataldo and Herbsleb (2010) describe transparency as "the open availability of any type of information regarding the development process from the design documents to the source code itself, development tasks, defects and interactions among the ecosystem's members." In relation to participation in the context of smart environments, particularly automated decision systems (ADS), Shadowen, Lodato, and Loi (2020) define transparency as "the ability of stakeholders to understand, audit, and/or impact a system."

2.2 Overview

This chapter provides a range of perspectives in Section 3 on transparency and the ambient in smart cities based on the research and practice literature, enabling formulation of a conceptual framework for operationalization and use in the exploration under investigation in this chapter. Section 4 provides a description of the methodology used in this chapter followed by the results and a discussion of findings. Section 5 identifies opportunities associated with future directions for research and practice and Section 6 concludes with chapter coverage highlights, the major contributions, limitations, and key insights and ideas. Several exercise questions are also included for the reader in support of further exploration.

The primary audiences for this chapter include students, educators, researchers, policy makers, businesses, community members, urban practitioners, and anyone concerned with transparency and the ambient in relation to visibilities and invisibilities in urban design, experiences, and opportunities in smart cities and regions.

3. PERSPECTIVES ON TRANSPARENCY AND THE AMBIENT IN SMART CITIES

In the context of visibilities and invisibilities in smart cities, this chapter provides a review of the research literature for transparency and then for the ambient.

3.1 Transparency and Smart Cities

Transparency is considered in relation to smart cities in this review of the research and practice literature in terms of data, design, openness, people, and technology.

Data

In an interview with the IoT Council (2011), Crutzen discusses the work of Langheinrich (2005) on data collection and privacy and observes that "without transparency humans will be without informational rights" while noting that in the case of Ambient Intelligence (AmI) "the invisibility of the data collection is not a mistake but intentional." Mulder (2015) highlights opportunities associated with an open(ed) mindset in relation to co-creative partnerships and more creative ways of working with data, sharing, and change in support of "a more transparent sociable city." Böhlen (2019) identifies the need to "make transparent what drives our data-driven decisions" that inform ambient intelligence (AmI) as an aid to understanding and testability. Hobson and Burgers (2019) describe the Society 5.0 initiative in Japan that "would blend together human and non-human interaction" contributing to "the integration of physical and cyber realms" involving "the 'internet of things', artificial intelligence, and related technologies." Hobson and Burgers (2019) urge caution, claiming that, "big data may be a boon for efficiency" yet "difficult questions related to privacy, responsibility, and transparency" as well as agency emerge, providing opportunities for "much more transparent accounting of the possible downsides surrounding this transition" while offering a space for "further consideration of how some of the risks can be mitigated." Lin, Crabtree, Dillo, et al. (2020) describe digital repository trustworthiness in terms of five guiding principles, one of which is transparency. It could be argued that the other four elements of the TRUST principle (responsibility, user focus, sustainability, and technology) are also highly relevant to smart cities. Davies (2020) provides a mapping of "the relationship between transparency, data use, standards and data infrastructure" as a way of "re-asserting the role of citizens within smart cities."

Design

Smalley Bowen (2007) describes transparency in relation to glass in the design of a structure "that seems to disappear" yet, and ironically, "the coatings typically used to enhance the energy performance of glazing make glass more difficult to recycle." From an architectural perspective, Morello and Piga (2015) focus on "people-centered design" with an "increased interest in transparency and participation in the process of decision making about urban spaces." In the context of art and architecture, Herzog and de Meuron (2016) employ transparencies through the use of glass and mirrors and other media. It is worth noting that, regarding privacy issues, Cavoukian and Popa (2016) advanced a privacy by design (PbD) framework based on seven principles guiding the embedding or coding of privacy preferences into technologies, the Internet of Things (IoT), or other emerging technologies, one of which is visibility and transparency. Husain, Hasan, Shafie, Hamidon, and Pandey

(2018) provide a review of transparent solar photovoltaic technologies and associated challenges. Nur-E-Alam and Yao (2019) describe transparency efforts in relation to solar windows for smart cities while Solar Mag (2020) points to new technology developments in the form of transparent solar panels "set to transform the global solar energy landscape." Stein, Draper, and Hellyer (2019) discuss the "nearly invisible" in the context of the "structural elements" of "glass structures" in the city of New York in relation to innovation opportunities. Shadowen et al. (2020) advance "a provisional framework for ADS transparency and accountability based on participative practices" for city-making, namely, "participatory governance and participatory design" where ADS refers to automated decision systems and encompasses the notion of algorithmic transparency to overcome issues such as "inscrutable, invisible pieces of code" citing the work of Eubanks (2017).

Openness

Gil-Garcia et al. (2016) identify openness among a series of dimensions for smartness in government, arguing that, "when a government is open, it becomes more transparent and more accountable" enabling "smarter government for its citizens, businesses, and other stakeholders" who are then able to "use information in smarter ways." Matheus and Janssen (2019) provide a systematic review of the literature for public administration (PA) and information systems (IS) to investigate transparency in the context of open government data. Finding that PA focuses more on participation and trust whereas IS focuses on user interface, experience, and data quality, Matheus and Janssen (2019) use Window Theory "to unify existing models" through "integrating a broad range of factors within a single model" because of the capability of Window Theory "to develop context-dependent models that are both comprehensive and parsimonious." Matheus and Janssen (2019) claim that Window Theory is "not a tested theory" but rather, "a comprehensive account of the determinants and effects of transparency." Although "the characteristics of glass let it transmit, reflect and refract light", Matheus and Janssen (2019) note that glass in windows for example "blurs the view", and as such, "many transparency efforts only reveal a part of the full picture." It is perhaps worth noting that Panetta (2019) identifies "transparency and traceability" requiring a focus on six elements, one of which is openness, as part of Gartner's top 10 strategic trends for 2020. Gil-Garcia, Gasco-Hernandez, and Pardo (2020) advance a framework in support of the "multidimensionality of open government as a socio-technical phenomenon" identifying five dimensions from the literature, one of which is transparency.

People

It is also worth noting that Gil-Garcia et al. (2016) indicate that from the research literature, "people, education, learning, and knowledge have been identified as of central importance to cities' ability to innovate" and the importance of learning city, learning government, and "transparent learning environments" for citizen engagement. For Matheus and Janssen (2019), "use is a central construct influencing transparency" highlighting the importance of "participation and interaction amongst people." Burns and Andrucki (2020) take issue with "the predominant hype around transparency" claiming that such terminology "relies on expectations of unpaid labor" in relation to smart city related events such as "hackathons" and organization-led initiatives such as Code for America, Codefest Marathon, Random Hacks of Kindness and the like. De Guimarães, Severo, Júnior, Da Costa, and Salmoria (2020) explore smart governance and quality of life (QoL) where several factors are measured in relation to QoL, one of which is transparency. In exploring what is said to be an "understudied issue", Desouza, Hunter, Jacob, and Yigitcanlar (2020) describe their research on "pathways to smart cities" identifying the three pathways of greenfield development, neighborhood development, and platform-oriented development. Desouza et al. (2020) note that "to the degree that 'transparency' is an objective of public sector service provision, the platform-pathway far exceeds the two development pathways" where in one platform example, "the community could contribute and become part of the discussion." Leorke (2020) reminds us of the notion that playful initiatives contribute to the "image" of the smart city as "transparent" and the importance of play for addressing complex urban issues and policies. In the context of quality of life and technologies in smart cities, Lai, Lai, and Lai (2021) point to the importance of "the right balance of protection, privacy, and transparency."

Technology

Brauneis and Goodman (2018) explore "what meaningful 'algorithmic transparency' entails" for smart cities from the perspective of law and technology and "present eight principal types of information that records concerning publicly implemented algorithms should contain." Visibility in the context of computer supported cooperative work (CSCW) and socio-technical systems and affordances, is, according to Crowston, Saltz, Rezgui, and Hegde (2019) "closely related to the concepts of awareness and of system transparency." Streitz et al. (2019) discuss the notions of "transparency and fairness" and "testable transparency" while highlighting the importance of "transparency and traceability of intelligent systems and their algorithms" now and going forward. Hakak, Khan, Gilkar, Imran, and Guizani (2020) address technologies for smart cities to improve the quality of life for people, such as the Internet of Things (IoT), cloud computing, and blockchain with a focus on security. Hakak et al. (2020) propose "a conceptual architecture for securing smart city using blockchain technology." According to Hakak et al. (2020) the 'transaction' concept in the context of smart cities and blockchain technologies "is verified data/information stored in a blockchain" and "smart-city-based applications require transparent transactions" as well as "no single point of failure, automatic decision making, and data protection to ensure the integrity and authorization of transactions" in addressing privacy and trust challenges, to name a few. Ismagilova, Hughes, Rana, and Dwivedi (2020) provide a review of the research literature on security, privacy, and risk in smart cities pointing to "threats relating to information security and challenges for smart city infrastructure in the management and processing of personal data" pertaining to "mobile devices and services, smart city infrastructure, power systems, healthcare, frameworks, algorithms and protocols to improve security and privacy, operational threats for smart cities, use and adoption of smart services by citizens, use of blockchain, and use of social media." Going forward, Ismagilova et al. (2020) identify the need for "a globally interconnected, transparent network of smart cities." Khan, Yaqoob, Tran, Kazmi, Dang, and Hong (2020) address edge-computing-enabled smart cities where features such as transparency in the use of blockchain technologies contribute to the potential for "secure and transparent sharing of data by citizens" in support of smart applications.

3.2 The Ambient and Smart Cities

Böhlen and Frei (2010) note that, "ambient technologies often seem to counter the urban qualities of event, flow and temporality" so as to "create an almost contradictory sense of consistency and coherency" whereas "the urban world that is experienced is one of change, flow and surprise." According to Böhlen and Frei (2010) "ambient systems design needs to come out from the bunker and move into full view" so that "it becomes a public affair" that crosses disciplines in order for "(AmI) engineers, managers and politicians" to "build the new ambient cities" where AmI refers to ambient intelligence. Rickert (2013) claims that, "an ambient rhetoric continually attunes itself both to what is present and to what withdraws" referring to an "age of ambience" in which "boundaries between subject and object, human and non-human, and information and matter dissolve." McCullough (2013) suggests that, "when you feel renewed sensibility to your surroundings you might try calling this *ambient*." Mattern (2014) refers to the notion of ambient interfaces and ambient data referring the reader to the work of McCullough (2013). Architecturally, Morello and Piga (2015) identify "increasing interest in simulation by society and the profession" in terms of "how simulation technologies will enable us to pay more attention to the intangible values of the ambiance of spaces" because "spatial design affects human well-being and health,"

Panetta (2019) speaks of the "ambient experience" in the context of "multiexperience" as one of Gartner's "top 10 strategic technology trends" where "people-literate technology" enables "interaction to include multisensory and multitouchpoint interfaces like wearables and advanced computer sensors." If, as Orasch (2019) claims, "the smart city exists all around us, in various forms and in various spaces" then, it may be that the smart city could be considered to be ambient. Streitz et al. (2019) articulate the ambient as ambient intelligence (AmI) involving "an approach that moves beyond 'smart-only' environments towards humane and sociable" in "redefining the 'smart-everything' paradigm" in support of "a human-/people-/citizen-centered design approach" keeping people "in the loop" and optimizing people's smartness in smart spaces. McCullough (2020) uses the ambient in relation to interactivity and "how context informs good design." Abhari and Abhari (2020) explore factors affecting the adoption of ambient intelligence (AmI) applications in architecture "enabled by recent advancements in artificial intelligence (AI) and Internet of Things (IoT) technologies" that "offer new mechanisms for data–driven decision making in architectural design." Yigitcanlar, Desouza, Butler, and Roozkhosh (2020) provide a review of the literature identifying challenges and risks for AI while calling for "a degree of transparency, adaptability (to respond to varying environmental conditions), and accountability (for levels of performance)."

3.3 Summary

Table 1 provides an overview by year and author for transparency and the ambient in the context of visibilities and invisibilities in smart cities and regions covering the timespan of 2007 to 2021.

In summary, over this multi-year period, transparency emerges in relation to glass (Smalley Bowen, 2007); software ecosystems (Cataldo and Herbsleb, 2010), people centered design (Morello & Piga, 2015), an open(ed) mindset (Mulder, 2015), openness in government (Gil-Garcia et al., 2016; Gil-Garcia et al., 2020), art and architecture using glass and mirrors (Herzog and de Meuron, 2016), algorithms from a law and technology perspective (Brauneis and Goodman, 2018), solar photovoltaic technologies (Husain et al., 2018), CSCW (Crowston et al., 2019), open government data (Matheus & Janssen, 2019), solar windows (Nur-E-Alam and Yao, 2019), openness and traceability (Panetta, 2019), glass structures (Stein et al., 2019), openness (Stephanidis et al., 2019), fairness / testability / traceability of transparency systems and algorithms (Streitz et al., 2019), security for society 5.0 (Hobson & Burgers, 2019), labor and the invisibility of digital care workers (Burns and Andrucki, 2020), mapping the relationship between transparency and data (Davies, 2020), measure of quality of life in relation to governance (De Guimarães et al., 2020), platforms as a pathway to smart cities (Desouza et al., 2020), blockchain for securing smart cities (Hakak, 2020), security, privacy, and risk in moving toward globally interconnected smart cities (Ismagilova et al., 2020), edge-computing-enabled smart cities (Khan et al., 2020), playfulness (Leorke, 2020), data repository trustworthiness (Lin et al., 2020), automated decision systems (Shadowen et al., 2020), transparent solar panels (Solar Mag., 2020), AI systems (Yigitcanlar et al., 2020), and quality of life and smart city data (Lai et al., 2021).

The ambient emerges in relation to technologies and experiences (Böhlen and Frei, 2010), the Commons (McCullough, 2013), rhetoric (Rickert (2013), interfaces and data (Mattern, 2014), ambience of spaces (Morello and Piga, 2015), multiexperience (Panetta, 2019), AmI / design context / smart societies (Streitz et al., 2019), application adoption in architecture (Abhari and Abhari, 2020), and interactivity, context and good design (McCullough, 2020).

Across this array of characteristics for transparency and the ambient, intersections emerge for data and openness appears multiple times in relation to transparency. Also of note is the presence of artificial intelligence (AI), ambient intelligence (AmI), and multiexperience.

Table 1. Overview of key elements characterizing transparency and the ambient in smart cities

Overview of Transparency and the Ambient in Smart Cities			
Author(s)	**Year**	**Transparency**	**The Ambient**
Smalley Bowen	2007	Glass	
Böhlen & Frei	2010		Technologies + experiences
Cataldo and Herbsleb	2010	Software ecosystems	
McCullough	2013		Commons
Rickert	2013		Rhetoric
Mattern	2014		Interfaces and data
Morello & Piga	2015	People centered design	Ambiance of spaces
Mulder	2015	Open(ed) mindset	
Gil-Garcia et al.	2016	Openness in government	
Herzong & de Meuron	2016	Art & Architecture	
Brauneis & Goodman	2018	Algorithmic	
Husain et al.	2018	Solar photovoltaic technologies	
Crowston et al.	2019	CSCW	
Matheus & Janssen	2019	Open government data	
Nur-E-Alam & Yao	2019	Solar windows	
Panetta	2019	Openness	Multiexperience
Stein et al.	2019	Glass structures	
Stephanidis et al.	2019	Openness	
Streitz et al.	2019	Fairness/Testability/Traceability	AmI; Design Context; Smart Societies
Hobson & Burgers	2019	Security in 'Society 5.0'	
Abhari & Abhari	2020		Application adoption in architecture
Burns & Andrucki	2020	Expectations of unpaid labor	
Davies	2020	Mapping the data relationship	
De Guimarães et al.	2020	Quality of life & Governance	
Desouza et al.	2020	Platform pathway to SCs	
Gil-Garcia et al.	2020	Open government	
Hakak et al.	2020	Blockchain for securing SCs	
Ismagilova et al.	2020	Network of global SCs	
Khan et al.	2020	Edge-computing-enabled SCs	
Leorke	2020	Play	
Lin et al.	2020	Data repository trustworthiness	
McCullough	2020		Interactivity, Context, and Good design
Shadowen et al.	2020	Automated decision systems	
Solar Magazine	2020	Solar panels	
Yigitcanlar et al.	2020	AI systems	
Lai et al.	2021	Data & SCs	

Figure 1. Transparency and the Ambient: Emerging aspects and domain coverage for smart cities

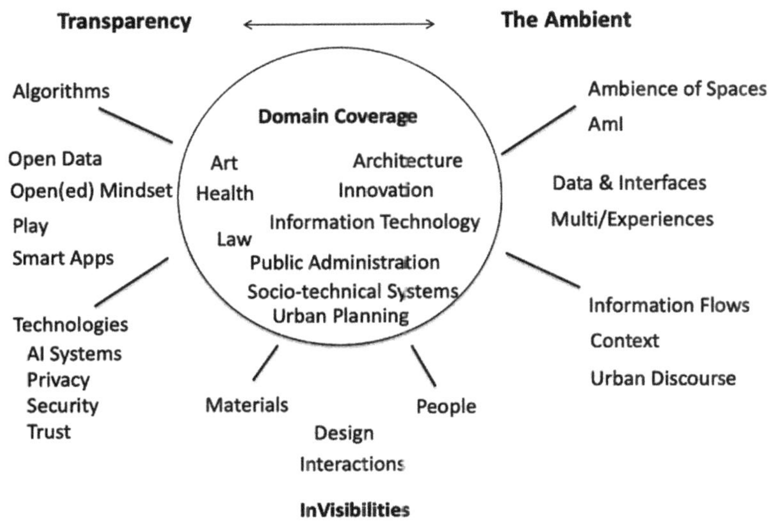

Graphically, emerging aspects of transparency and the ambient in smart cities are depicted in Fig. 1 based on this review of the literature, providing a glimpse of the domain coverage, from art to law, to public administration, to name a few.

3.4 Conceptualizing Transparency and the Ambient in Smart Cities

Based on the review of the research literature in this chapter, a conceptual framework is developed and operationalized for transparency and the ambient in smart cities and regions. As depicted in Fig. 2, in the context of visibilities and invisibilities in smart cities, openness supports transparency while attuning to urban spaces supports the ambient, providing opportunities for research and practice.

This framework guides exploration of the research question for this chapter presented in Section 2 and reformulated here as a proposition, as follows.

P1: Insight into the relationship between transparency and the ambient in smart cities sheds light on the many dimensions of visibilities and invisibilities through the use of openness and attuning to urban spaces along with associated complexities and opportunities for research and practice.

Figure 2. Framework for transparency and the ambient in smart cities

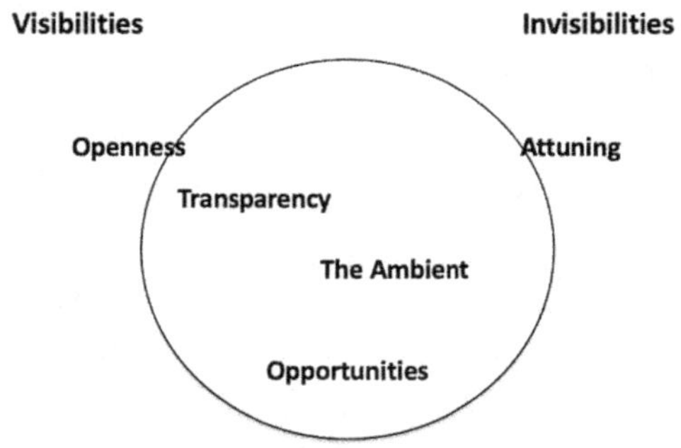

4. METHODOLOGY FOR EXPLORING TRANSPARENCY AND THE AMBIENT

Guided by the range of smart city perspectives explored through the research literature in this chapter, transparency and the ambient are investigated as dimensions of urban visibilities and invisibilities. For the purposes of this exploration, *openness* is used as a proxy for *transparency* and *attuning to urban spaces* is used as a proxy for *the ambient*.

4.1 Approaches for Exploring Transparency and the Ambient in Smart Cities

An exploratory case study approach is used in this chapter combined with an explanatory correlational design for investigating transparency and the ambient in smart cities and regions. The study underlying this chapter was described using a website and during sign up demographic data were collected including elements such as age range, location (city), and people could self identify in one or more categories (e.g., community member, city official, researcher, educator, student, business, etc.). Upon sign up, people were invited to complete a survey about their experience of their city as a smart city. An invitation was also provided to participate in an in-depth interview about smart cities, online or in person, depending upon the location. The study attracted participation from across Canada and extended for example to cities in the United States, Ireland, Greece, and Israel. The survey instrument, containing a mix of closed and open-ended questions, and the interview protocol were pretested prior to use in the study. To enable responses from a broad spectrum of people, heterogeneity sampling, a type of purposive sampling was used whereby an invitation to participate in the study was posted to webspaces with the potential to attract a wide response. Data analysis included the use of content analysis for qualitative data and descriptive statistics for quantitative data. The Real Statistics Software Resource Pack Add In (Zaiontz, 2019) for Microsoft Excel was used, consisting as it does of a wide range of tools for statistical analysis, such as the correlation feature for calculation of the Spearman correlation for ordinal data, generated by assessments from survey respondents. The survey instrument and interview protocol enable people to assess their experience of the city and region where they live in terms of a range of elements pertaining to transparency and the ambient. Data were also collected in parallel with this study through group and individual discussions with a wide range of people across many sectors in small to medium to large-sized cities in Canada (e.g., greater Victoria, Vancouver, and Toronto). Overall, an analysis of data consisted of n=78 including 41% females and 59% males for people ranging in age from their 20s to their 70s across multiple continents (e.g., Europe, North America, and the Middle East).

4.2 Transparency and the Ambient in Smart Cities – An Exploration

Qualitatively, city information technology (IT) staff indicated that, regarding urban data, "fundamentally there is a desire to be very, very open" since "it is public data we manage on behalf of the citizenry" and "maybe it helps improve something." City IT staff added that, "regarding physical infrastructure" opportunities arise in that, "how we interpret using it is still open." An educator in higher education indicated that "we'll open our doors and we can have different age groups" creating an "interplay between communities and postsecondary institutions." Another educator spoke in terms of "freedom and openness" in support of a 'Wise City' in order "to find the way to include both" as in, wisdom and smartness for "the IoT, big data, and other tech systems to make the city a better place."

Regarding attuning to urban spaces, when asked, "What enables you to feel the pulse of the city?" a community leader and member responded with, "the city's amenities: parks, transit, kiosks, social media, events, and retail (dining especially)." An educator responded to this question about the pulse of the city by saying, "the multilayers of senses" along with "the feeling of dynamic" while a student stated, "human activity in every corner." A community member and leader highlighted the importance of "smart cities" as the "chief catalyst for the changes that are needed in cities" at this time, adding that, "primary efforts should be focused around people and their needs" suggestive of attuning, "with everything else playing a supporting role."

Quantitatively, when asked to assess to what extent smart cities are associated with openness, people responded to several items for this survey question using a 7-point Likert-type scale (e.g., 1 = Not at all, 2 = Not sure, 3 = Maybe, 4 = Neutral, 5 = Sort of, 6 = Sure, 7 = Absolutely). For example, Table 2 shows assessments for the item *openness* as a proxy for *transparency*. For *openness*, 17% of respondents selected the neutral position of 4 on the scale, 33% selected 6 on the scale and 50% selected 7 on the scale. When asked to assess whether city-focused social media and other aware technologies give rise to many possibilities, people responded to several items for this survey question using a 7-point Likert-type scale (e.g., 1 = Not at all, 2 = Not sure, 3 = Maybe, 4 = Neutral, 5 = Sort of, 6 = Sure, 7 = Absolutely).

Table 2. Assessments for Transparency (openness) and the Ambient (attuning) in smart cities

Items	Assessments	Correlation
Transparency (openness)	17% (4); 33% (6); 50% (7)	.43
The Ambient (attuning to urban spaces)	17% (4); 33% (6); 50% (7)	

For example, Table 2 shows assessments for the item *attuning to urban spaces* as a proxy for *the ambient*. For *attuning to urban spaces*, 17% of respondents selected position 4 on the scale, 33% selected position 6 on the scale and 50% selected 7 on the scale.

This assessment enabled the correlating of *transparency* and *the ambient* using the proxies of *openness* and *attuning to urban spaces* where, using the Real Statistics Software (Zaiontz, 2019) Resource Pack add-in for Microsoft Excel, which contains a variety of statistical functions and data analysis tools, the correlation feature was used to calculate the Spearman correlation, resulting in a correlation of .43.

Looking at the scatterplot for this relationship, as openness (transparency) increases or decreases, so too does attuning (the ambient), as depicted in Fig. 3.

Figure 3. Scatterplot: transparency (openness) and the Ambient (attuning to urban spaces) in smart cities

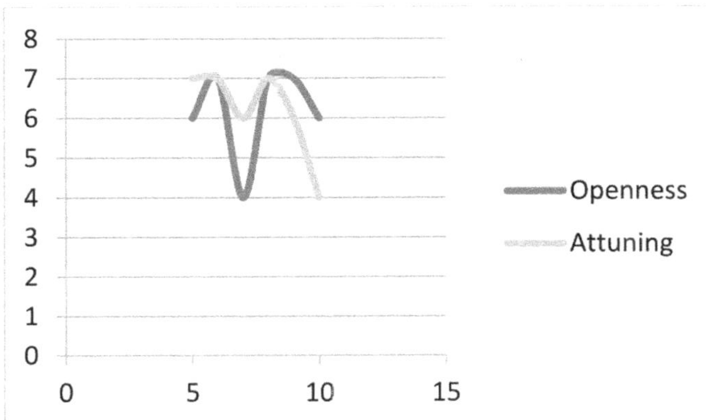

4.3 Discussion of the Value of Transparency and the Ambient in Smart Cities

Transparency in the form of *openness* emerged in this work in several ways including through data, education, and collaborations. The importance of *attuning to urban spaces* in relation to "people and their needs" was voiced as a "primary focus" for smart cities as an articulation of the ambient. The correlation of .43 achieved in this work, based on assessments of *transparency (openness)* and *the ambient (attuning to urban spaces)* in smart cities, falls in the range of .35 to .65, where according to Creswell (2018), "correlations above .35 are useful for limited prediction" and are, "typical values used to identify variable membership in the statistical procedure of factor analysis." While some prediction potential emerges between *transparency* and *the ambient* in this chapter, using the proxies of *openness* and *attuning to urban spaces*, opportunities arise for learning more about smart cities through such perspectives and approaches and this is discussed in Section 5. It is worth noting the openness mindset on the part of city IT staff in relation to the use of public data and the potential that "maybe it helps improve something" and extending to interpretations for the use of physical infrastructure. What is key here is support for the notion by Matheus and Janssen (2019) of "use" through participation and interaction, also voiced by Streitz et al. (2019) in terms of "keeping people in the loop" in "optimizing people's smartness." Where Matheus and Janssen (2019) note the incompleteness of many transparency efforts and the need for Window Theory to offer a unifying and integrating of multiple factors, this work advances the importance of *the ambient* using *attuning to urban spaces* as a proxy to provide additional factors of interest to urban researchers and practitioners.

Additionally, the presence of neutral responses for both *openness* and *attuning to urban spaces* emerged and it is worth noting that Nadler, Weston, and Voyles (2015) explored the use and interpretation of such mid-points. Fakhrhosseini and Jeon (2017) for example, explore emotion/affect induction methods (e.g., imagination) in relation to human-computer interaction (HCI) where "neutral" emerges as one of eight emotional states, building on earlier work by Gross and Levenson (1995).

5. EMERGING PRACTICE AND RESEARCH OPPORTUNITIES

Future and emerging opportunities for practice and research are discussed in Sections 5.1 and 5.2 pertaining to transparency and the ambient in smart cities and regions.

5.1 Practice

In practical terms, this chapter contributes to opportunities associated with understandings of *transparency* and *the ambient* in smart cities using the proxies of *openness* and *attuning to urban spaces*.

Transparency (Openness): Exposure to, and examples of *transparency* are provided through the proxy of *openness* for understanding smart city visibilities and invisibilities. As such, this chapter points to emerging potentials for the use of data, design, and experiences in relation to openness. Crucially then, this work highlights the need for involving people more meaningfully in the development of initiatives for urban problem-solving guided by more open approaches and mindsets (Mulder, 2015).

The Ambient (Attuning to Urban Spaces): This chapter surfaces understandings of *the ambient* through the proxy of *attuning to urban spaces,* contributing to experiences associated with urban data, design, and privacy. As such, this work highlights the need for more meaningful participation in the development of initiatives for urban problem-solving that are more dynamic, adaptive and responsive (in keeping with characterizations of the ambient by Stephanidis et al., 2019) to contemporary urban opportunities, and not without challenges, in the context of emergent smart and future cities.

5.2 Research

This chapter advances the potential for future research opportunities and emerging discourse spaces for *transparency* (based on explorations of *openness)* and of *the ambient* (in the form of *attuning to urban spaces)* in the context of smart cities and regions.

Transparency (Openness): This work reinforces the importance of effects pertaining to transparency initiatives (Matheus & Janssen, 2019) while being attentive to the meaningful involvement of people in support of enabling better understanding of the use potentials and the value of urban data.

The Ambient (Attuning to Urban Spaces): Conceptualization of spaces for real-time, dynamic, and adaptive types of interactions in this chapter opens the way for possibly novel considerations of more aware people interacting with each other and with increasingly awareness-enabled technologies.

As such, areas for exploration potential for the interplay of *transparency* and *the ambient* include:

1. *Data* initiatives focusing on improving the quality of the urban experience
2. *Design* focused around "people and their needs"
3. *Experiences* fostering interactions that accommodate both wisdom and smartness

Additionally, this work directs attention to the potential for explorations of the viability of the conceptual framework for transparency and the ambient in smart cities, operationalized for use in this chapter in demonstrating early-stage potentials, while offering and opening up evolving urban spaces for testing and practice.

6. CONCLUSION

This chapter provides an overview of the dimensions of *transparency* and *the ambient* in smart cities through an exploration of selected aspects of *openness* and *attuning to urban spaces* in the context of visibilities and invisibilities. Data, design, and experiences emerge as key elements associated with openness and attuning to urban spaces for navigating emerging interaction potentials in smart cities. These elements will be enabled through fostering additional education and opportunities for involving people more meaningfully with aware technologies. The main contributions of this chapter include: a) articulation and operationalization of a conceptual framework for transparency and the ambient in smart cities; b) arguing for a re-thinking and extending of dimensions for visibilities and invisibilities to include *transparency* and *the ambient* represented in this work by the proxies of *openness* and *attuning to urban spaces*; and c) highlighting the elements of data, design, and experiences in fostering smarter, more responsive, cities and regions.

Limitations of this chapter pertain to the small sample size of the study upon which this exploration is based. Limitations of this chapter related to the use of proxies for transparency (e.g., openness) and the ambient (e.g., attuning to urban spaces) are mitigated by actual examples from the research literature and in-depth discussions with a range of individuals regarding everyday emergent practices. Other limitations related to literature review domain coverage are mitigated by the interdisciplinary potential to extend this review to other areas going forward.

This chapter continues to give form, detail, and definition to the exploration of dimensions of invisibilities and visibilities in smart cities and regions. The analysis of findings identifies a relationship between transparency and the ambient, having "limited prediction" potential, with a positive correlation of .43, yet such correlations are said (Creswell, 2018) to be important for the exploration of the "interconnection of variables" going forward, relevant to the study of spaces for smart cities. Where Gil-Garcia et al. (2020) reflect on dimensions of open government in terms of moving beyond transparency, this chapter opens the way for considerations of translucence in the chapter that follows.

Insights of particular interest to the reader include the following:

1. The emerging range of dimensions for invisibilities and visibilities in the form of *transparency* and *the ambient* in smart cities
2. The presence of a "limited prediction" potential emerged when *openness* (as a proxy for transparency) and *attuning to urban spaces* (as a proxy for the ambient) were correlated in a smart cities context
3. The interplay of data, design, and experiences associated with *transparency* and *the ambient*

Ideas highlighted in this chapter pertain to the following:

1. The importance of people becoming more involved in understandings of *openness* and *attuning to urban spaces* in relation to *transparency*

and *the ambient* as dimensions of invisibilities and visibilities in smart cities
2. The potential for people to shed light on *transparency* and *the ambient* in urban spaces
3. Explorations of *transparency* and *the ambient* as novel practice and research spaces for smarter cities and regions

Exercises for the reader to consider, encouraging further exploration, include the following:

1. Identify one or more examples of *openness* and *transparency* in your city
2. Identify one or more examples of *attuning to urban spaces* and *the ambient* in your city
3. Describe why *transparency* and *the ambient* could be important for smart cities, based on your example(s)

If you would like to share your responses to the Exercises with the author, please complete the questions in this online space for Chapter 3 – Transparency and the Ambient

This work will be of interest to a broad audience including students, educators, researchers, artists, urban planners, urban practitioners, community members, and anyone concerned with emerging understandings of spaces for transparency (openness) and the ambient (attuning to urban spaces) along with associated opportunities for renderings of the invisible and visible in smart, responsive, and future cities.

REFERENCES

Abhari, M., & Abhari, K. (2020). Ambient intelligence applications in architecture: Factors affecting adoption decisions. In K. Arai, S. Kapoor, & R. Bhatia (Eds.), *Advances in Information and Communication. FICC 2020. Advances in Intelligent Systems and Computing* (Vol. 1129). Springer. doi:10.1007/978-3-030-39445-5_18

Böhlen, M. (2019). A.I. ethics in the city. In M. Antona & C. Stephanidis (Eds.), Lecture Notes in Computer Science: Vol. 11572. *Universal Access in Human-Computer Interaction. Theory, Methods and Tools. HCII 2019*. Springer. doi:10.1007/978-3-030-23560-4_2

Böhlen, M., & Frei, H. (2010). Ambient intelligence in the city: Overview and new perspectives. In H. Nakashima, H. Aghajan, & J. C. Augusto (Eds.), *Handbook of ambient intelligence and smart environments* (pp. 911–938). Springer. doi:10.1007/978-0-387-93808-0_34

Brauneis, R., & Goodman, E. P. (2018). Algorithmic transparency for the smart city. *The Yale Journal of Law of Technology*, *20*, 103–176.

Cataldo, M., & Herbsleb, J. D. (2010). *Architecting in software ecosystems: Interface translucence as an enabler for sociable collaboration. In ECSA 2010*. ACM. doi:10.1145/1842752.1842772

Cavoukian, A., & Popa, C. (2016). *Embedding privacy into what's next: Privacy by design for the Internet of Things*. Ryerson University, Privacy & Big Data Institute. Retrieved 29 April 2018 from https://www.ryerson.ca/content/dam/pbdce/papers/Privacy-by-Design-for-the-Internet-of-Things.pdf

Creswell, J. W. (2018). *Educational Research: Planning, Conducting, and Evaluating Quantitative and Qualitative Research* (6th ed.). Pearson.

Crowston, K., Saltz, J. S., Rezgui, A., & Hegde, Y. (2019). Socio-technical affordances for stigmergic coordination implemented in MIDST, a tool for data-science teams. *Proc. ACM Hum.-Comput. Interact.*, 3, 1-25. 10.1145/3359219

Davies, T. (2020). Shaping participatory public data infrastructure in the smart city. In K. S. Willis & A. Aurigi (Eds.), *The Routledge companion to smart cities* (pp. 74–89). Routledge. doi:10.4324/9781315178387-7

De Guimarães, J. C. F., Severo, E. A., Júnior, L. A. F., Da Costa, W. P. L. B., & Salmoria, F. T. (2020). Governance and quality of life in smart cities: Towards sustainable development goals. *Journal of Cleaner Production*, *253*, 119926. doi:10.1016/j.jclepro.2019.119926

Desouza, K. C., Hunter, M., Jacob, B., & Yigitcanlar, T. (2020). Pathways to the making of prosperous smart cities: An exploratory study on the best practice. *Journal of Urban Technology*, *27*(3), 3–32. doi:10.1080/10630732.2020.1807251

Eubanks, V. (2017). *Automating Inequality: How High-Tech Tools Profile, Police, and Punish the Poor*. St. Martin's Press.

Fakhrhosseini, S. M., & Jeon, M. (2017). Affect/emotion induction methods. In *Emotions and Affect in Human Factors and Human-Computer Interaction* (pp. 235–253). Elsevier. doi:10.1016/B978-0-12-801851-4.00010-0

Gil-Garcia, J. R., Gasco-Hernandez, M., & Pardo, T. A. (2020). Beyond transparency, participation, and collaboration? A reflection on the dimensions of open government. *Public Performance & Management Review*, *43*(3), 483–502. Advance online publication. doi:10.1080/15309576.2020.1734726

Gil-Garcia, J. R., Zhang, J., & Puron-Cid, G. (2016). Conceptualizing smartness in government: An integrative and multi-dimensional view. *Government Information Quarterly*, *33*(3), 524–534. doi:10.1016/j.giq.2016.03.002

Gross, J. J., & Levenson, R. W. (1995). Emotion elicitation using films. *Cognition and Emotion*, *1995*(9), 87–108. doi:10.1080/02699939508408966

Hakak, S., Khan, W. Z., Gilkar, G. A., Imran, M., & Guizani, N. (2020). Securing smart cities through blockchain technology: Architecture, requirements, and challenges. *IEEE Network*, *34*(1), 8–14. doi:10.1109/MNET.001.1900178

Herzog, J., & de Meuron, P. (2016). *Treacherous transparencies: Thoughts and observations*. Actar Publishers.

Hobson, C., & Burgers, T. (2019). Security in 'Society 5.0'. *Japan Times*. Retrieved 23 June 2020 from https://www.japantimes.co.jp/opinion/2019/06/05/commentary/japan-commentary/security-society-5-0/#.XvIkB1KZM-c

Husain, A. A. F., Hasan, W. Z. W., Shafie, S., Hamidon, M. N., & Pandey, S. S. (2018). A review of transparent solar photovoltaic technologies. *Renewable & Sustainable Energy Reviews*, *94*, 779–791. doi:10.1016/j.rser.2018.06.031

IoT Council. (2011). *Council interview with Cecile Crutzen on ambient intelligence and Internet of Things*. Retrieved 23 January 2020 from https://www.theinternetofthings.eu/content/council-interview-cecile-crutzen-ambient-intelligence-and-internet-things

Ismagilova, E., Hughes, L., Rana, N. P., & Dwivedi, Y. K. (2020). Security, privacy and risks within smart cities: Literature review and development of a smart city interaction framework. *Information Systems Frontiers*, 1–22. doi:10.100710796-020-10044-1 PMID:32837262

Khan, L. U., Yaqoob, I., Tran, N. H., Kazmi, S. A., Dang, T. N., & Hong, C. S. (2020). Edge-Computing-Enabled Smart Cities: A Comprehensive Survey. *IEEE Internet of Things Journal, 7*(10), 10200–10232. doi:10.1109/JIOT.2020.2987070

Lai, C. S., Lai, L. L., & Lai, Q. H. (2021). Smart city. In *Smart Grids and Big Data Analytics for Smart Cities* (pp. 1–171). Springer. doi:10.1007/978-3-030-52155-4_1

Langheinrich, M. (2005). *Personal privacy in ubiquitous computing tools and support systems*. Dissertation of the Swiss Federal Institute of Technology, Zürich. Retrieved 23 January 2020 from http://www.vs.inf.ethz.ch/publ/papers/langheinrich-phd-2005.pdf

Leorke, D. (2020). Reappropriating, reconfiguring and augmenting the smart city through play. In A. Nijholt (Ed.), *Making Smart Cities More Playable: Exploring Playable Cities*. Springer. doi:10.1007/978-981-13-9765-3_3

Lin, D., Crabtree, J., Dillo, I., Downs, R. R., Edmunds, R., Giaretta, D., De Giusti, M., L'Hours, H., Hugo, W., Jenkyns, R., Khodiyar, V., Martone, M. E., Mokrane, M., Navale, V., Petters, J., Sierman, B., Sokolova, D. V., Stockhause, M., & Westbrook, J. (2020). The TRUST principles for digital repositories. *Scientific Data, 7*(1), 144. doi:10.103841597-020-0486-7 PMID:32409645

Lytras, M. D., & Visvizi, A. (2018). Who uses smart city services and what to make of it: Toward interdisciplinary smart cities research. *Sustainability, 10*(6), 1998. doi:10.3390u10061998

Matheus, R., & Janssen, M. (2019). A systematic literature study to unravel transparency enabled by open government data: The window theory. *Public Performance & Management Review*. Advance online publication. doi:10.1080/15309576.2019.1691025

Mattern, S. (2014). Interfacing urban intelligence. *Places Journal*. Retrieved 1 February 2020 from https://placesjournal.org/article/interfacing-urban-intelligence/

Mattern, S. (2020). Post-It Note City. *Places Journal*. Retrieved 16 February 2020 from https://placesjournal.org/article/post-it-note-city/

McCullough, M. (2013). *Ambient commons: Attention in the age of embodied information*. The MIT Press., doi:10.7551/mitpress/8947.001.0001

McLeod, J., & Zochodne, G. (2020). Sidewalk Labs pulls out of Toronto smart city project after 3 years, citing 'unprecedented economic uncertainty'. *Financial Post.* Retrieved 22 April 2021 from https://financialpost.com/technology/sidewalk-labs-pulls-out-of-toronto-quayside-smart-city-project-because-of-unprecedented-economic-uncertainty

Morello, E., & Piga, B. E. A. (2015). Experiential simulation in architecture and urban space: Introduction to the special issue. *Ambiances: International Journal of Sensory Environment, Architecture and Space*, 671.

Mulder, I. (2015). Opening up: Towards a sociable smart city. In *Citizen's Right to the Digital City*. Springer. doi:10.1007/978-981-287-919-6_9

Nadler, J. T., Weston, R., & Voyles, E. C. (2015). Stuck in the middle: The use and interpretation of mid-points in items on questionnaires. *The Journal of General Psychology*, *142*(2), 71–89. doi:10.1080/00221309.2014.994590 PMID:25832738

Nur-E-Alam, M., & Yao, A. (2019). Transparent solar windows: From labs to industry, towards smart cities. In *Encyclopedia*. MDPI.

Orasch, A. K. (2019). Seeing the smart city: Mapping technologies in Canada. *The Conversation*. Retrieved 19 December 2019 from https://theconversation.com/seeing-the-smart-city-mapping-technologies-in-canada-112458

Panetta, K. (2019). Gartner top 10 strategic technology trends for 2020. *Smarter With Gartner*. Retrieved 7 March 2020 from https://www.gartner.com/smarterwithgartner/gartner-top-10-strategic-technology-trends-for-2020/

Rickert, T. (2013). *Ambient rhetoric: The attunements of rhetorical being*. University of Pittsburgh Press., doi:10.2307/j.ctt5hjqwx

Sáinz-Piña, R. M. (2011). *Smart cities: A first step towards the Internet of Things*. Fundación Telefónica.

Scholl, H. J., & AlAwadhi, S. (2016). Creating smart governance: The key to radical ICT over-haul at the City of Munich. *Information Polity*, *21*(1), 21–42. doi:10.3233/IP-150369

Shadowen, N., Lodato, T., & Loi, D. (2020). Participatory governance in smart cities: Future scenarios and opportunities. In N. Streitz & S. Konomi (Eds.), Lecture Notes in Computer Science: Vol. 12203. *Distributed, Ambient and Pervasive Interactions. HCII 2020* (pp. 443–463). Springer. doi:10.1007/978-3-030-50344-4_32

Smalley Bowen, T. (2007). Glass: Transparent, translucent, and ironic. *Architectural Record, 15*(October). Retrieved June 12, 2020, from https://www.architecturalrecord.com/articles/4011-glass-transparent-translucent-and-ironic

Solar Mag. (2020). Transparent solar panels: Reforming future energy supply. *Solar Magazine*. Retrieved 9 June 2020 from https://solarmagazine.com/solar-panels/transparent-solar-panels/

Stein, M., Draper, P., & Hellyer, R. (2019). City of glass: Recent advancements in glass structures in New York City. *Structural Engineering International, 29*(1), 101–111. doi:10.1080/10168664.2018.1519365

Stephanidis, C., Salvendy, G., Antona, M., Chen, J. Y. C., Dong, J., Duffy, V. G., Fang, X., Fidopiastis, C., Fragomeni, G., Fu, L. P., Guo, Y., Harris, D., Ioannou, A., Jeong, K. K., Konomi, S., Krömker, H., Kurosu, M., Lewis, J. R., Marcus, A., ... Zhou, J. (2019). Seven HCI Grand Challenges. *International Journal of Human-Computer Interaction, 35*(14), 1229–1269. doi:10.1080/10447318.2019.1619259

Streitz, N., Charitos, D., Kaptein, M., & Böhlen, M. (2019). Grand challenges for ambient intelligence and implications for design contexts and smart societies. *Journal of Ambient Intelligence and Smart Environments, 11*(1), 87–107. doi:10.3233/AIS-180507

Townsend, A. M. (2013). *Smart cities: Big data, civic hackers and the quest for a new utopia*. W. W. Norton & Company.

Yigitcanlar, T., Desouza, K. C., Butler, L., & Roozkhosh, F. (2020). Contributions and risks of artificial intelligence (AI) in building smarter cities: Insights from a systematic review of the Literature. *Energies, 2020*(13), 1473. doi:10.3390/en13061473

Zaiontz, C. (2019). *Real Statistics using Excel*. www.real-statistics.com

ADDITIONAL READING

Del Bello, L. (2020). How COVID-19 could redesign our world. *BBC Future*. Retrieved 9 June 2020 from https://www.bbc.com/future/article/20200527-coronavirus-how-covid-19-could-redesign-our-world

Dutta, M., Reecti, R., & Mandal, S. K. (2018). Light transmitting and self cleansing concrete as a smart building material: A review. *International Conference on Smart City and Emerging Technology (ICSCET)*. 10.1109/ICSCET.2018.8537328

Erkartal, P. O., & Uzunkaya, A. (2019). Transparency as a component of public space. *IOP Conference Series. Materials Science and Engineering*, *471*(9), 092026. doi:10.1088/1757-899X/471/9/092026

Hackett, R. (2020). Upgrade, interrupted: Alphabet's Sidewalk Labs signed on to build the city of the future. Then the "techlash" kicked in. *Fortune*.

Lisdorf, A. (2020). The 7 forces of artificial intelligence in cities. *Meeting of the Minds blog*. Retrieved 20 April 2020 from https://meetingoftheminds.org/the-7-forces-of-artificial-intelligence-in-cities-33053

Watts, J. (2020). Blue-sky thinking: How cities can keep air clean after coronavirus. *The Guardian*. Retrieved 7 June 2020 from https://www.theguardian.com/environment/2020/jun/07/blue-sky-thinking-how-cities-can-keep-air-clean-after-coronavirus

Wharton. (2019). What's fueling the smart city backlash? *Knowledge@Wharton Blog*. Philadelphia: University of Pennsylvania. Retrieved 10 June 2020 from https://knowledge.wharton.upenn.edu/article/whats-behind-backlash-smart-cities/

Xu, H., & Geng, X. (2019). People-centric service intelligence for smart cities. *Smart Cities*, *2*(2), 135–152. doi:10.3390martcities2020010

KEY TERMS AND DEFINITIONS

Ambient: In the context of smart cities, the ambient describes the increasing presence of more aware people together with awareness-enabled technologies in support of interactivity, influencing human communication and activities, human-machine interactivities, and machine-machine interactivities and in turn affecting behavior, experience, information capabilities, economies, literacies, and indeed everything.

Attuning: Attuning refers to the concept or quality of being attuned and aware as it applies to people on the one hand, to technologies on the other, within physical and digital spaces and networks.

Openness: Openness in this work refers to an approach or mindset for working with others designing, and the like, and is used in this chapter as a proxy for transparency.

Smart Cities: Smart cities are urban areas and regions characterized by more aware and engaged people, interacting with each other and aided by the use of more awareness enhancing technologies for a wide range of purposes from communication to design to atmospheres to mobility to livability, to governance, to data generation and use for learning and informed decision-making.

Transparency: Transparency refers to greater ease in seeing or making visible what is occurring in relation to the physical, digital, and less tangible.

Chapter 4
Translucence and the Ambient:
Emergent Perspectives in Smart Cities

ABSTRACT

The purpose of this chapter is to explore translucence and the ambient as dimensions of urban visibilities and invisibilities in the context of emergent perspectives on smart cities to enrich understandings. The research literature for translucence and the ambient is explored in terms of urban visibilities and invisibilities in smart cities and regions, enabling the formulation and operationalization of a conceptual framework to guide the investigation in this chapter. Using an exploratory case study approach, combined with an explanatory correlational design, survey and interview responses provide a mix of quantitative and qualitative data for analysis. This chapter makes a contribution to 1) the research literature for translucence and the ambient in smart cities and regions, 2) the surfacing of additional urban elements for understanding 21st century cities, and 3) the evolving of theory through formulation of a conceptual framework for translucence and the ambient in smart cities.

1. INTRODUCTION

Where Gil-Garcia, Gasco-Hernandez, and Pardo (2020) point to the possible need for looking "beyond transparency, participation, and collaboration",

DOI: 10.4018/978-1-7998-3850-0.ch004

this chapter is motivated to consider additional dimensions of visibilities and invisibilities in smart cities through an exploration of *translucence* and *the ambient* in a variety of ways and spaces. From an interdisciplinary perspective, translucence is explored in this chapter from the social (Erickson and Kellogg, 2000) to learning (Prestigiacomo, Hadgraft, Hunter, and Locker, 2020) to interfaces (Cataldo and Herbsleb, 2010) to materials in urban spaces and regions such as wood (Boitouzet, 2017), concrete (Dutta, Reecti, and Mandal, 2018), glass (Stein, Draper, and Hellyer, 2019), and UltraSurfaces (Ferrara and Pasetti, 2020). Regarding the ambient, Streitz, Charitos, Kaptein, and Böhlen (2019) describe ambient intelligence (AmI) as supportive of the humane and social in smart environments where people and their smartness is enabled with "an approach that moves beyond 'smart-only' environments" in "redefining the 'smart-everything' paradigm." Guiding this investigation is the use of an exploratory case study along with an explanatory correlational design where data are collected through the use of multiple methods including survey and in-depth interviews. In parallel with this study, data are also systematically gathered through individual and group discussions. This chapter is significant in that it explores the nature of the relationship between *translucence* and *the ambient* using the proxies of *visible infrastructures* and *awareness*, respectively. This chapter makes a contribution through a) extending the research literature for invisibilities and visibilities in smart cities to include translucence and the ambient; b) the evolving of perspectives on smart cities and regions; and c) formulation of a conceptual framework for translucence and the ambient in smart cities. As such, the objective of this work is described below.

Objective: The objective of this chapter is to explore and advance emergent understandings of translucence and the ambient in smart cities and regions. The key research question posed is – *What is the nature of the relationship between translucence and the ambient in smart cities and regions, as dimensions of urban visibilities and invisibilities?*

2. BACKGROUND AND OVERVIEW

This chapter addresses the physical/material, the social, and other emergent forms of translucence in the context of smart cities and regions. For example, Erickson and Kellogg (2000) articulated the notion of social translucence in systems design in support of the social dimension of processes. Erickson and Kellogg (2000) explain that they "speak of socially *translucent* systems rather than socially *transparent* systems" due to the "vital tension between privacy and visibility." Additionally, Erickson and Kellogg (2000) claim that, "translucence stands in more generally for the power of constraints" and it is worth noting that smart cities are constrained in many ways, not least of which is that of financial constraints (Nisenbaum, 2019). In response to concerns with the environment, Boitouzet (2017) articulates the notion of "translucent wood" as a low-carbon option for cities of the future while Dutta, Reecti, and Mandal (2018) describe the idea of translucent concrete.

2.1 Definitions

Definitions are presented here, based on the research literature, for key terms used in this chapter. Additional terms with definitions, as articulated by this work, are included in the *Key Terms and Definitions* section at the end of this chapter.

Ambient: Streitz, Charitos, Kaptein, and Böhlen (2019) claim that the ambient, in the context of ambient intelligence (AmI), "was proposed to contrast" developments that were "technology-driven" by "paying attention to user-centered design, social interfaces and the notion of context-aware and adaptive ambient environments."

Social Translucence: Erickson and Kellogg (2000) describe social translucence from the perspective of "an approach to designing digital systems" as "making social information visible within the system."

Translucence: In describing conceptualizations of interface translucence, Cataldo and Herbsleb (2010) note that, "the goal is to increase visibility of some particular pieces of information or behaviors while hiding others."

2.2 Overview

This chapter provides a variety of perspectives in Section 3 on transparency and the ambient in smart cities through a review of the research and practice literature. A conceptual framework is then developed and operationalized for use in the exploration conducted in Section 4. Section 4 also includes a description of the methodology used in the exploration followed by the results and a discussion of findings. Section 5 describes opportunities associated with future directions for research and practice and Section 6 concludes with chapter coverage highlights, the major contributions, limitations, and key insights and ideas. Several exercise questions are also included for the reader in support of further exploration.

The primary audiences for this chapter include students, educators, researchers, policy makers,-businesses, community members, and urban practitioners concerned with extending spaces for working with visibilities and invisibilities associated with urban design, urban understandings and representations, and other opportunities in the context of smart cities, responsive cities, and future cities.

3. PERSPECTIVES ON TRANSLUCENCE AND THE AMBIENT IN SMART CITIES

In the context of visibilities and invisibilities, this chapter provides a review of the research and practice literature for translucence and smart cities; a review of the literature for translucence and the ambient in smart cities and regions; and formulation of a conceptual framework for translucence and the ambient in smart cities.

3.1 Translucence and Smart Cities

Translucence is considered in relation to smart cities in this review in terms of social translucence; interface translucence; building materials; and art, discovery, and learning.

Social Translucence

Advancing the notion of "socially translucent systems" characterized by "visibility, awareness, and accountability," Erickson and Kellogg (2000) claim that, "constraints act on visibility" noting that, "our ability to see things decays as we move farther away from them" whereas "social software" enables "greater resolution to people nearby." On the other hand, Erickson and Kellogg (2000) state that, "awareness arises from visibility." According to Erickson and Kellogg (2000), socially translucent systems support graceful "communication and collaboration among large groups of people over computer networks" through the idea of technologies with windows instead of walls, so as to "mesh with human behavior" at "individual and collective levels." This translucence, afforded by windows, allows "users to 'see' one another, to make inferences about the activities of others, to imitate one another," in support of emerging "new social forms." Gilbert (2012) argues that, "people want social translucence, but network structure gets in the way" in the context of "modern social media." In response, Gilbert (2012) claims to have "built theory relating social translucence to social network structure, updating social translucence for the era of social networks." McDonald, Gokhman, and Zachry (2012) describe social translucence as "a stance toward the design of systems that allows users to better understand collaborative system participation through awareness of contributions and interactions." Barretto, Szóstek, and Karapanos (2013) advance a model for the design of socially translucent systems in relation to behavior change where "visibility, awareness, and accountability" are elements said to be needed for effective communication.

Interface Translucence

Weiser and Seely Brown (1995) employed a window metaphor for technologies that support visibility, enabling the viewing of information, people, and much more. In the context of interface translucence, Cataldo and Herbsleb (2010) utilize the "three key informational elements" of "uncertainty, complexity, and activity awareness" arguing that interface translucence "constrains transparency by emphasizing relevant pieces of information for both technical and socio-organizational purposes" such that "information overload is limited and sensitive information is protected." In the context of urban spaces, Lehtovuori (2010) refers to the "confusion" that "resides in the idea of visibility" and the notion of "metaphorical visibility" where "the idea that space is intelligible and whatever there is in space can be known, assumes that space is translucent." DiLonardo (2020) reminds us that translucence in urban elements is not new and has existed in the form of glass prisms in sidewalks in many cities around the world including Seattle, San Francisco, New York, Chicago, Philadelphia, London, Dublin, Melbourne, Sydney, and Amsterdam, to name a few, and extending to smaller cities as well. According to DiLonardo (2020), the functional and artistic pieces of glass embedded in sidewalks are also known as "vault lights" or "sometimes called pavement lights in the UK" and designed "to allow light into the basement areas below ground" with the first vault light patent dating back to 1834 by Edward Rockwell. DiLonardo (2020) notes that vault lights provided "a way to illuminate spaces where natural light wasn't available and a way to avoid using glass, oil, and candles" with a decline in use "by the 1930s when electricity became more common and inexpensive." Ringuette

(2007) describes the translucent glass featured in a number of sidewalks in Victoria, British Columbia as "historic sidewalk prisms." The thousands of "three-inch square purple glass prisms" were designed "in the early days of Victoria" to provide light "when the afternoon sun streams through the old glass" and "the dark under-sidewalk tunnel is transformed" (Ringuette, 2007). Ringuette (2007) explains that, "most downtown basements extended to the street curb" and "spaces under sidewalks provided extra storage for merchants and access from outside the building to deliver coal and freight" referred to as "areaways" in Vancouver (Hagemoen, 2018). Of the "surviving 11,155 intact prisms", people "appreciate those prisms every day from the street level" explains Ringuette (2007). DiLonardo (2020) identifies work on the "history and value" of lights such as that by Wong (2011).

Building Materials

Smalley Bowen (2007) describes the translucency of glass in "story-tall light monitors" of a museum structure. Gehl (2010) articulates the notion of translucence in terms of "being able to see through windows" as important for people in the city, from inside and outside a building. Brunetti, Chartsia, Ros, and Ingrassia (2016) advance the notion of translucent societies as a way for cities to accommodate the migrations that are occurring around the world. Brunetti et al. (2016) propose "an infrastructure, a hub which is set on top of the existing public or private buildings." Brunetti et al. (2016) add that the benefits for existing residents of the structure "is energy through solar panels and it changes the cityscape adding a new layer which stands as a sculpture on the terraces." Brenna (2016) describes a "translucent city bus" that is "powered by wind." Dutta et al. (2018) describe ceiling translucence in relation to "light transmitting and self cleansing concrete" employed "as a smart building material." Translucent facades for buildings (Materials, 2019), are said to be designed "to hide unsightly building elements and assist in protecting users from harmful UV rays, while also ensuring maximum thermal conduction." Jewell (2018) describes "translucent 'hugging' towers" designed to possibly aid in cleaning the air pollution in Hong Kong. Wang (2019) highlights an auditorium construction "wrapped in a translucent skin of ETFE panels that floods the interior with natural light" in daytime and "glows like a lantern when illuminated from within at night" contributing to an "ethereal feel" where ETFE refers to Ethylene Tetra Fluoro Ethylene, "a fluorine-plastic fabric."

Art, Discovery, and Learning

Lewis (2009) refers to New York as "the capital of glass, the city of windows" as in, "transparent, translucent, and opaque; reflective, tinted, frosted, coated, clear" adding that "the slightest shift in the angle of sun fall can hide or reveal entire worlds." For Lewis (2009), "glass remains a mystery" providing "an endless inspiration to architects and designers as it grows stronger, lighter, clearer, and more flexible." Steiner (2011) describes the renovation of a classic historic building, the Hamilton City Hall, as an example of a civic space incorporating translucence. Among the many interior additions, and "the most contentious" according to Steiner (2011), is "a new translucent glass-walled boardroom" that is situated "in the middle of the wide hall" on the second floor and "is visible from the public lobby" and "surrounded by the councilors' offices." Noting that, "city leaders often look to cutting edge landmarks as a way to bolster their city's image" another addition to this urban infrastructure is "a 'luminous ceiling' composed of backlit translucent panels running the length of the second floor" that "passes over the boardroom box below." Morgan (2011) argues that "places and journeys which provide an encounter with 'otherness', whether in terms of different cultures or the 'more than human realm' of nature and wilderness, have significant transformative potential" in support of learning. As such, Morgan (2011) describes the notion of "Thin Places" from "Celtic Christian tradition", where "the 'veil' between the worlds was particularly 'thin' and 'translucent'" enabling "communion" with landscape elements from trees to mountains, to boundaries and temples.

Fetterman, Weber, Freking, Volpe, and Scott (2014) describe the LuminoCity project as, "a novel three-dimensional data display" where "a 3D printed model of Cambridge, MA was generated from LADAR data." Thus, through laser detection and ranging (LADAR), a dataset was developed enabling "a translucent plastic model" to be "cast from a mold of the 3D printed model" and "a display system" was then developed "to project data onto the translucent model" enabling the projection of "a wide range of analyses onto the city" that includes "satellite imagery and network traffic." In the context of art, Gavish (Times Hudson Valley, 2018) presents translucent images (photographic and painted) as part of "crystalline architecture" that "incorporates photo-transfers and uses watercolor and ink to create and build cities that expand in vertical and horizontal rows." Referring to the versatility of wood, Cross (2018) describes the work of "a team of scientists at the University of Maryland" who "developed a process that can make wood stronger and more durable – and also translucent" making it ideal as a building material while being "a good thermal insulator." In the context of learning, using a human-centered approach, Prestigiacomo et al., (2020) focus on the three principles of translucence, namely, visibility, awareness, and accountability where visibility pertains to "the capability of seeing relevant information that enhances people's understanding of the activity under scrutiny"; awareness "entails a well-informed understanding of a visible situation"; and accountability "is about the implications of sharing educational data on teachers' or learners' responsibility". Gardner (2020) highlights translucent building materials used in Japanese architecture including textiles, aluminum mesh skin, and glass, to name a few.

3.2 Translucence and the Ambient in Smart Cities

Translucence and the ambient is considered in this review of the research and practice literature in relation to social translucence, healthcare, and ambient experience and ambient computing.

Social Translucence

Dadlani et al. (2014) discuss social translucence in terms of how it "provides a much needed handle on social interaction issues relevant for ambient intelligence systems."

Healthcare

Kaiying, Plewe, and Röcker (2015) explore the notion of "an Asian approach to ambient systems" in relation to smart technologies and "hypermultitasking" where "emotional support" is provided "by increasing the assurance users can have in their ability to survey and monitor" environmental elements such as air quality and more generally, "the invisible toxic surroundings." Cook, Duncan, Sprint, and Fritz (2018) propose the use of information and communication technologies (ICTs) in smart cities in making healthcare smarter by employing ambient sensors, described as, "sensors embedded in the environment." For example, Cook et al. (2018) argue that "the use of ambient intelligence technology" in smart cities "may improve quality of life" for residents "but also save the city substantial costs of patient treatment." Cook et al. (2018) are also attentive to the challenges of ambient sensors in smart cities pertaining to privacy and security mitigated in part by "privacy-preserving data mining techniques." Additionally, Cook et al. (2018) are attentive to accessibility and usability and to "connection with other services" such that "weaving sensors into existing city features" serves to "help researchers understand the connection between city design and health."

Ambient Experience and Ambient Computing

Bizzocchi (2018) describes his work with "the genre", ambient video, as being inspired by the work of Brian Eno with ambient music and video. Ambient video according to Bizzocchi (2018) is not new, pointing to the "Christmas TV Yule Log" available on many television channels dating back to the 1960s, the "Shaw Cable Frame channel" and many others. Bizzocchi (2018) works with ambient video art, incorporating the "large, higher-resolution television frames" for the "presentation of beautiful moving image visuals-like fine-art photography, but acting as a living window into a world outside our daily routine." Having "adapted" the dictum by Eno "that an ambient medium must be as easy to ignore as it is to notice", Bizzocchi (2018) maintains that ambient video art is "slow paced" and "is resolutely non-narrative" since "narrative is fundamentally inconsistent with ambient experience." For Bizzocchi (2018), three "rules" guide his notion of ambient video – "never require your attention; always reward your attention with visual pleasure; and sustain its visual pleasure over repeated viewing." Panetta (2019) describes multiexperience as one of several strategic trends identified by Gartner, encompassing "multisensory and multitouchpoint" becoming what will be understood as "an ambient experience." Daosabah, Guermah, and Nassar, (2019) describe ambient computing as "being interested in adapting services to the needs of the user" where "a service is independent of all software platforms or material and can be invoked by clients or other services." Celentano and Dubois (2019) refer to ambient computing where "the perception of the locations fades into a continuum space surrounding the user."

3.3 Summary

Table 1 provides an overview by year and author for translucence and the ambient in the context of visibilities and invisibilities in smart cities and regions covering the timespan of 2000 to 2020.

Table 1. Overview of key elements characterizing translucence and the ambient in smart cities

Overview of Translucence and the Ambient in Smart Cities			
Author(s)	**Year**	**Translucence**	**The Ambient**
Erickson & Kellogg	2000	Social	
Smalley Bowen	2007	Glass	
Ringuette	2007	Glass in sidewalk prisms	
Lewis	2009	Glass	
Böhlen & Frei	2010		Technologies + experiences
Cataldo & Herbsleb	2010	Activity awareness, Complexity, Uncertainty	
Lehtovuori	2010	Urban spaces	
Gehl	2010	Building materials (glass/windows)	
Morgan	2011	Learning	
Steiner	2011	Renovations in historical building	
Gilbert	2012	Social	
McDonald et al.	2012	Social	
Dadlani et al.	2014	Social	Social interaction issues
Fetterman et al.	2014	Urban data displays	
Kaiying et al.	2015		Hypermultitasking
Brenna	2016	City bus powered by wind	
Brunetti et al.	2016	Societies to accommodate refugees	
Boitouzet	2017	Building materials (wood)	
Bizzocchi	2018		Video; Experiences
Cook et al.	2018		Sensors + healthcare - privacy & security - accessibility & usability - city design & health
Cross	2018	Building materials (wood)	
Dutta et al.	2018	Building materials (concrete)	
Gavish (TimesHV)	2018	Crystalline architecture	
Jewell	2018	Green Skyscrapers	
Materials	2019	Facades	
Panetta	2019		Experiences
Wang	2019	Auditorium – ETFE panels	
DiLonardo	2020	Vault lights and sidewalk glass prisms	
Gardner	2020	Building materials in Japanese architecture	
Prestigiacomo et al.	2020	Accountability, Awareness, Visibility	

In summary, over a 20 year period, translucence appears in relation to glass (Smalley Bowen, 2007; Ringuette, 2007; Lewis, 2009; Dilonardo, 2020); the social (Erickson and Kellogg, 2000; McDonald et al., 2012; Gilbert, 2012; Dadlani et al., 2014); renovations in historical buildings (Steiner, 2011); activity awareness, complexity, and certainty (Cataldo and Herbsleb, 2010); urban spaces (Lehtovuori, 2010); learning (Morgan, 2011); urban data display (Fetterman et al., 2014); a translucent city bus powered by wind (Brenna, 2016); translucent societies accommodating refugees and nomadic populations (Brunetti et al., 2016); building materials (Gardner, 2020; Cross, 2018; Dutta et al., 2018; Boitouzet, 2017; Gehl, 2010); crystalline architecture (Gavish in Times Hudson Valley, 2018); urban towers (Jewell, 2018); facades (Materials, 2019); ETFE (Ethylene tetrafluoroethylene) panels for a conference center auditorium (Wang, 2019); and accountability, awareness, and visibility (Prestigiacomo et al., 2020).

The ambient emerges in relation to technologies and experiences (Böhlen & Frei, 2010); social interaction issues (Dadlani et al., 2014); video and ambient experience (Bizzocchi, 2018); the assurance enabled by technologies for hypermultitasking in monitoring and informing users about ambient elements such as air quality (Kaiying et al., 2015); sensors and healthcare (Cook et al., 2018); and experiences (Panetta, 2019).

Figure 1. Translucence and the Ambient: Emerging aspects and domain coverage for smart cities

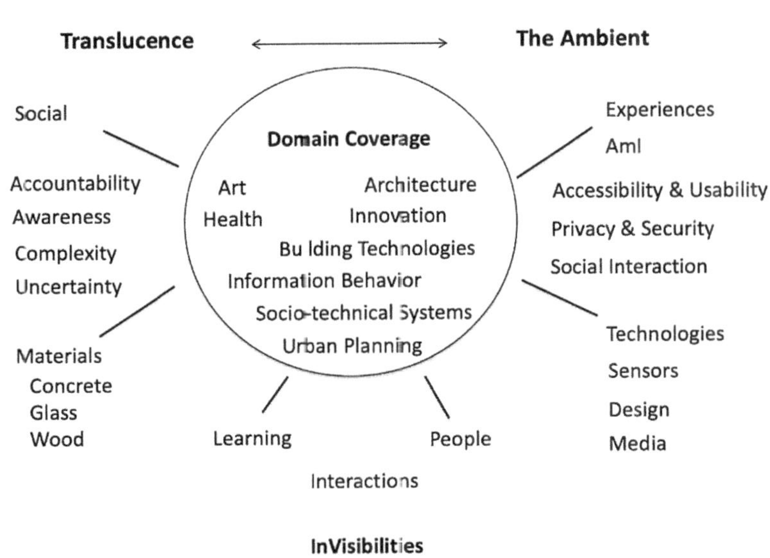

Across this array of characteristics for translucence and the ambient, intersections emerge for the social and the social and awareness emerge multiple times in relation to translucence while experiences emerge multiple times for the ambient. Also of note is the presence of complexity, uncertainty, and sensors. In the context of visibilities and invisibilities in smart cities, infrastructural elements support translucence while awareness supports both translucence and the ambient, providing opportunities for research and practice.

Graphically, emerging aspects of translucence and the ambient in smart cities are depicted in Fig. 1 based on this review of the literature, providing a glimpse of the domain coverage, from art to architecture, to health, to building materials, to technologies, to experiences, to name a few.

3.4 Conceptualizing Translucence and the Ambient in Smart Cities

Based on the review of the research literature in this chapter, a conceptual framework for translucence and the ambient in smart cities is advanced as depicted in Figure 2.

In the context of visibilities and invisibilities in smart cities, *visible infrastructures* in the form of emerging materials, among other elements, in urban spaces support understandings of *translucence* while *awareness* supports *the ambient*, providing opportunities for research and practice.

Figure 2. Framework for translucence and the ambient in smart cities

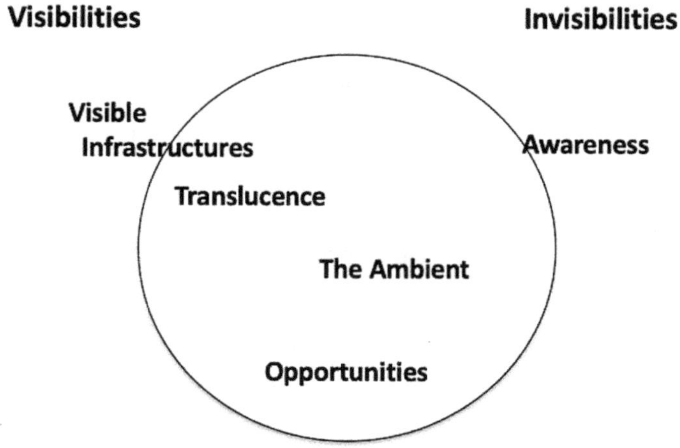

This framework contributes further to guidance for exploration of the research question for this chapter presented in Section 1 and reformulated here as a proposition, in the following paragraph.

P1: Understanding the relationship between transparency and the ambient in smart cities enriches the dimensionalities of urban visibilities and invisibilities while highlighting associated complexities and opportunities.

4. METHODOLOGY FOR EXPLORING TRANSLUCENCE AND THE AMBIENT

This chapter employs a case study approach along with an explanatory correlational design in exploring translucence and the ambient in smart cities and regions. The proxies of *visible infrastructures* and *awareness* are used in this exploration for *translucence* and *the ambient*, respectively.

4.1 Research Design for Exploring Translucence and the Ambient in Smart Cities

This chapter employs a hybrid research design incorporating use of an exploratory case study approach as recommended by Yin (2018) for the study of contemporary phenomena, in this instance, transparency and the ambient in smart cities and regions. An explanatory correlational design is also used as it is said by Yin (2018) to accommodate "complementarity of case study and statistical research." Through use of a website, the study underlying the research for this chapter was described and during sign up demographic data were collected including elements such as age range, location (city), and people could self identify in one or more categories (e.g., community member, city official, researcher, educator, student, business, etc.). Upon sign up, people were invited to complete a survey about their experience of their city as a smart city. An invitation was also provided to participate in an in-depth interview about smart cities, online or in person, depending upon location. The study attracted participation from people in cities across Canada and extended to other cities in countries such as the United States, Ireland, Greece, and Israel.

The survey instrument, containing a mix of closed and open-ended questions, and the interview protocol were pretested prior to use in the study.

To enable response from a broad spectrum of people, heterogeneity sampling, a type of purposive sampling was used whereby an invitation to participate in the study was posted to webspaces with the potential to attract a wide response. Data analysis included the use of content analysis for qualitative data and descriptive statistics for quantitative data. The Real Statistics Software Resource Pack Add In (Zaiontz, 2019) for Microsoft Excel was used, consisting as it does of a wide range of tools for statistical analysis, such as the correlation feature for calculation of the Spearman correlation for the ordinal data, generated by assessments from survey respondents. The survey instrument and interview protocol were designed to enable people to assess their experience of the city and region where they live in terms of a wide variety of urban elements inclu awareness and visible infrastructures. Data were also collected in parallel with this study through group and individual discussions with a wide range of people across many sectors in small to medium to large-sized cities in Canada (e.g., greater Victoria, Vancouver, and Toronto). Overall, an analysis of data consisted of n=78 including 41% females and 59% males for people ranging in age from their 20s to their 70s.

4.2 Exploring Translucence and the Ambient in Smart Cities

When asked to assess to what extent elements such as visible infrastructures and awareness contribute to the making of smart cities on a scale of 1 (Not at all) to 7 (absolutely), as indicated in Table 2, 25% responded at position 5 on the 7-point Likert-type scale for visible infrastructures and 75% at the upper end of the scale at position 7. For awareness, 25% responded at position 5, 25% at position 6, and 50% at position 7. Quantitatively then, based on the correlation feature of the Real Statistics Software Add in for Microsoft Excel (Zaiontz, 2019), *translucence*, using the proxy *visible infrastructures* is correlated with *the ambient* using the proxy of *awareness* as ordinal data, showing the Spearman correlation coefficient result of .27.

Table 2. Assessments for Translucence and the Ambient in smart cities

Items	Assessments	Correlation (Spearman)	Correlation (Kendall)
Translucence (visible infrastructures)	25% (5); 75% (7)	.27	.25
The Ambient (awareness)	25% (5); 25% (6); 50% (7)		

According to Creswell (2018), for correlations in the .20 to .35 range, "there is only a slight relationship." It may be worth noting that in assessing wellness through the Internet of Things (IoT), McLeod, Spachos, and Plataniotis (2020) note that the Kendall correlation "measures the strength of dependency between the two variables" and in the case of transparency and the ambient, a Kendall correlation of .25 was found.

Looking at the scatterplot for this relationship, as *visible infrastructures* (*translucence*) increase or decrease, so too does *awareness* (*the ambient*), as depicted in Fig. 3.

Figure 3. Scatterplot for visible infrastructures (translucence) & awareness (the ambient) in smart cities

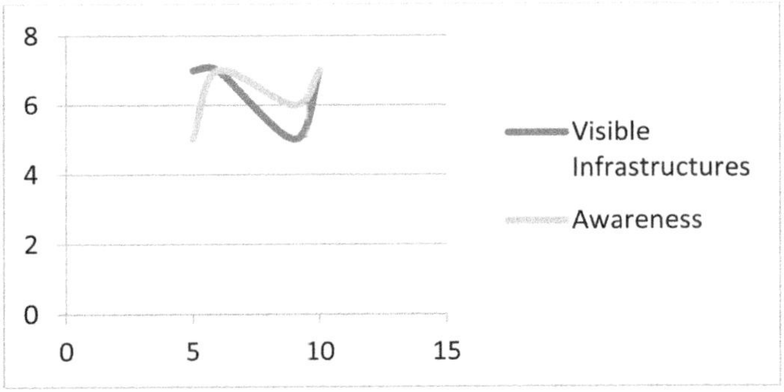

Qualitatively, when asked to comment on smart cities, an educator highlighted the importance of "community weaving, trust relationships between the authorities and the public" and the recognition that "each city finds its own way to become smart" since "no two smart cities look the same." In describing what smartness looks like, the educator pointed to the many ways in which visible physical infrastructures are being navigated and used through wi-fi, urban displays, and the "mobile phones" that "are everywhere." A community leader described smart cities as "those that are citizen-centric and inclusiveness-minded." While aware technologies were said, "to assist in citizen/visitor education and awareness," the community leader stated that, "unfortunately, most cities are not utilizing them in this manner." Responding to the question of how to get people more involved in the city and elements that would contribute to a smart city, the educator suggested a variety of approaches from "collaboration projects between public authorities and community groups where citizens are considered partners" to "appropriating the concept" of smart cities to "implementing those technologies that would have the highest impact in solving problems the city has." However, a community leader identified the need to "make involvement as intuitive and frictionless as possible" while being "focused around people and their needs."

4.3 Discussion and Interpretation of Findings

The exploration in this chapter reveals the presence of a "slight relationships" (Creswell, 2018) between *translucence (visible infrastructures)* and *the ambient (awareness)* in smart cities. From the perspective of an educator, the social dimension of smart cities emerges through reference to "community weaving, trust relationships between the authorities and the public." Additionally, the notion of constraints, connected as it is with translucence according to Erickson and Kellogg (2000), emerged through recognition that "each city finds its own way to become smart" since "no two smart cities look the same." The many ways in which visible physical infrastructures are being navigated and used were described through wi-fi, urban displays, and "mobile phones." Accountability, said to be one of several principles of translucence (Prestigiacomo et al., (2020), emerged in a number of ways pertaining to awareness and physical infrastructures from "collaboration projects between public authorities and community groups where citizens are considered partners" to "appropriating the concept" of smart cities to "implementing those technologies that would have the highest impact in solving problems the city has." If, as Panetta (2019) predicts, the "ambient experience" will encompass one of Gartner's "top 10 strategic technology trends" as the "multiexperience" then it could be argued that people and their experiences become critical in smart cities, now and going forward.

5. EMERGING PRACTICE AND RESEARCH OPPORTUNITIES

Emerging opportunities are identified in this chapter for practice and research in Sections 5.1 and 5.2 pertaining to translucence and the ambient in smart cities and regions.

5.1 Practice

In terms of practice, this chapter contributes to spaces for the rethinking of opportunities associated with the visible and invisible in smart cities based on understandings of *translucence* and *the ambient* using the proxies of *visible infrastructures* and *awareness*.

Translucence (Visible Infrastructures)

This chapter extends understandings of visible urban infrastructures in smart cities to include elements of translucence pointing to emerging potentials for the use of evolving materials in urban architecture, displays, and spaces; the importance of urban experiences; and interactions with aware technologies for education, awareness, and collaboration.

The Ambient (Awareness)

This chapter further elucidates the notion of *the ambient* associated with awareness in urban environments in relation to people and their experiences of urban environments and spaces. Crucially, this work invites participation by practitioners in the development of initiatives to shed light on urban problem-solving for developing more dynamic, adaptive and responsive solutions to contemporary urban challenges and opportunities.

5.2 Research

This chapter advances the potential for future research opportunities and emerging discourse spaces for *translucence* and *the ambient* in the context of smart cities.

Translucence and the Ambient

The viability of the conceptual framework for translucence and the ambient operationalized for use in this chapter demonstrates early-stage potentials while offering and opening up new understandings for evolving urban spaces for testing and practice.

Conceptualization of spaces for real-time, dynamic, and adaptive interactions emerge in this chapter, opening the way for new and possibly novel roles and collaborations for citizens, students, businesses, researchers, and governments. As such, areas for exploration potential include:

1. *Accountability* fostered through "trust relationships between the authorities and the public" and "where citizens are considered partners"
2. *Constraints* acknowledged by the unique set of circumstances guiding approaches, decisions, and choices in each city
3. *Social* fostered through community interactions and being more people-focused with an open and inclusive mindset

Additionally, this work directs attention to the potential for modification of existing materials and the appropriating of emerging materials for urban research in contributing to adaptive frameworks for smart, responsive, and future cities in support of more dynamic, interactive, and novel responses to solution-making in ever-more uncertain and complex environments.

7. CONCLUSION

This chapter opens spaces for consideration of translucence and the ambient as dimensions of visibilities and invisibilities for smart cities in relation to developments emerging from a range of disciplinary perspectives involving the innovating of materials such as wood, glass, and concrete for urban infrastructures. Accountability, constraints, and the social emerge as key elements for translucence and the ambient in the context of navigating invisibilities and visibilities in smart environments and socio-technical systems. These elements will be enabled through fostering increased education and adaptive thinking. The main contributions of this chapter include: a) articulation and operationalization of a conceptual framework for translucence and the ambient in the context of smart cities; b) further development of the research literature for visibilities and invisibilities in smart environments by arguing for a re-thinking and extending of dimensions to include *translucence* and *the ambient* represented in this work by the proxies of *visible infrastructures* and *awareness*; and c) highlighting the elements of accountability, constraints, and the social in fostering smarter, more responsive, cities and regions.-

Limitations of this chapter pertain to the small sample size of the study upon which this exploration is based. Additional limitations of this chapter related to the use of *visible infrastructures* and *awareness* as proxies for the exploration of *translucence* and *the ambient* could be mitigated going forward by exploring the latter more directly in future studies. Additional limitations related to literature review domain coverage are mitigated by the interdisciplinary potential to extend this review to other areas going forward.

This chapter continues to give form, detail, and definition to the exploration of dimensions of visibilities and invisibilities in smart cities and regions. The analysis of findings identifies a correlation of .27 between *translucence* and *the ambient*, constituting what Creswell (2018) describes as a "slight" relationship with the potential, going forward, "to explore the interconnection of variables" relevant to the study of such spaces for smart cities and regions.

Insights of particular interest to the reader include the following:

1. The emerging range of dimensions for invisibilities and visibilities in the form of *translucence* and *the ambient* in smart cities
2. The presence of a "slight" relationship emerged when *translucence* (e.g., *visible infrastructures)* and *the ambient* (e.g., *awareness)* were correlated in a smart cities context
3. The interplay of accountability, complexity, constraints, and the social associated with *translucence* and *the ambient*

Ideas highlighted in this chapter pertain to the following:

1. The importance of people becoming more involved in understandings of *translucence* and *the ambient* as dimensions of invisibilities and visibilities in smart cities
2. The potential for people to contribute to *translucence* and *the ambient* in urban spaces
3. Explorations of *translucence* and *the ambient* as novel practice and research spaces for smarter cities and regions

Exercises for the reader to consider, encouraging further exploration include the following:

1. Identify one or more examples of *translucence in visible infrastructures* in your city
2. Identify one or more examples of *awareness* and *the ambient* in your city
3. Describe why *translucence* and *the ambient* could be important for smart cities, based on your example(s)

If you would like to share your responses to the Exercises with the author, please complete the questions in this online space for Chapter 4 – Translucence and the Ambient

This work will attract the interest of a diverse audience including students, educators, researchers, artists, designers, urban planners, urban practitioners, community members, and anyone concerned with emerging spaces for *translucence* and *the ambient* that take into consideration the complex urban challenges and opportunities posed for renderings of the invisible and visible in smart, responsive, and future cities.

REFERENCES

Barreto, M., Szóstek, A., & Karapanos, E. (2013). An initial model for designing socially translucent systems for behavior change. *CHItaly '13, Proceedings of the Biannual Conference of the Italian Chapter of SIGCHI*. 10.1145/2499149.2499162

Bizzocchi, J. (2018). *Ambient landscapes*. Surrey, BC: Surrey Art Gallery. Retrieved 23 June 2020 from https://www.surrey.ca/files/BizzocchiSAGPresentsFinal.pdf

Böhlen, M., & Frei, H. (2010). Ambient intelligence in the city: Overview and new perspectives. In H. Nakashima, H. Aghajan, & J. C. Augusto (Eds.), *Handbook of ambient intelligence and smart environments* (pp. 911–938). Springer. doi:10.1007/978-0-387-93808-0_34

Boitouzet, T. (2017). *Woodoo: The augmented and translucent wood for the low-carbon cities of the future*. EDF. Retrieved 9 March 2020 from https://www.edf.fr/en/woodoo-augmented-wood

Brenna, L. (2016). Dancer bus, the translucent city bus that is powered by wind. *LifeGate*. Retrieved 12 June 2020 from https://www.lifegate.com/dancer-bus-powered-by-wind

Brunetti, F., Chartsia, M.-K., Ros, P., & Ingrassia, M. (2016). *Translucent societies | Encrypted nomadism*. Barcelona: Institute for Advance Architecture of Catalonia (IAAC). Retrieved 9 June 2020 from http://www.iaacblog.com/projects/translucent-societies-encrypted-nomadism/

Cataldo, M., & Herbsleb, J. D. (2010). *Architecting in software ecosystems: Interface translucence as an enabler for sociable collaboration. In ECSA 2010*. ACM. doi:10.1145/1842752.1842772

Celentano, A., & Dubois, E. (2019). An analytical frame for describing multi-* interactive systems. In *Proceedings of the 13th Bianual Conference of the Italian SIGCHI Chapter: Designing the next interaction (CHItaly '19)*. Association for Computing Machinery. 10.1145/3351995.3352040

Cook, D. J., Duncan, G., Sprint, G., & Fritz, R. L. (2018). Using smart city technology to make healthcare smarter. *Proceedings of the IEEE*, *104*(4), 708–722. doi:10.1109/JPROC.2017.2787688 PMID:29628528

Creswell, J. W. (2018). *Educational Research: Planning, Conducting, and Evaluating Quantitative and Qualitative Research* (6th ed.). Pearson.

Cross, D. T. (2018). Windows from transparent wood could replace glass. *Sustainability Times: Green News on Tech, Business, and the Environment*. Retrieved 22 June 2020 from https://www.sustainability-times.com/clean-cities/windows-from-transparent-wood-could-replace-glass/

Dadlani, P. (2014). SoPresent: An awareness system for connecting remote households. In *Ambient Intelligence: European Conference, AmI 2014*. Eindhoven, The Netherlands: Springer.

Daosabah, A., Guermah, H., & Nassar, M. (2019). Dynamic composition of services: An overview of approaches led by the context and intent of the user. In *Proceedings of the 4th International Conference on Big Data and Internet of Things (BDIoT '19)*. Association for Computing Machinery. 10.1145/3372938.3372960

DiLonardo, M. J. (2020). What those beautiful glass prisms in the sidewalk are really for. *Treehugger: Sustainability for All*. Retrieved 23 June 2020 from https://www.treehugger.com/sidewalk-vault-lights-pavement-lights-lluminate-underground-rooms-4869445

Dourish, P., & Bell, G. (2007). The infrastructure of experience and the experience of infrastructure: Meaning and structure in everyday encounters with space. *Environment and Planning. B, Planning & Design*, *34*(3), 414–443. doi:10.1068/b32035t

Dutta, M., Reecti, R., & Mandal, S. K. (2018). Light transmitting and self cleansing concrete as a smart building material: A review. *International Conference on Smart City and Emerging Technology (ICSCET)*. 10.1109/ICSCET.2018.8537328

Erickson, T., & Kellogg, W. A. (2000). Social translucence: An approach to designing systems that support social processes. *ACM Transactions on Computer-Human Interaction*, *7*(1), 59–83. doi:10.1145/344949.345004

Ferrara, M., & Pasetti, C. (2020). UltraSurfaces: A new material design vision. IHSI 2020, AISC 1131, 909-915. doi:10.1007/978-3-030-39512-4_139

Fetterman, M. R., Weber, Z. J., Freking, R. A., Volpe, A., & Scott, D. (2014). LuminoCity: A 3D printed, illuminated city generated from LADAR data, *IEEE International Conference on Technologies for Practical Robot Applications (TePRA)*, 1-4. 10.1109/TePRA.2014.6869161

Gardner, W. O. (2020). Liquid cities. *Places Journal*. Retrieved 22 June 2020 from https://placesjournal.org/article/liquid-cities/

Gehl, J. (2010). *Cities for people*. Island Press.

Gil-Garcia, J. R., Gasco-Hernandez, M., & Pardo, T. A. (2020). Beyond transparency, participation, and collaboration? A reflection on the dimensions of open government. *Public Performance & Management Review*, *43*(3), 483–502. Advance online publication. doi:10.1080/15309576.2020.1734726

Gilbert, E. (2012). Designing social translucence over social network. *CHI '12: Proceedings of the SIGCHI Conference on Human Factors in Computing Systems,* 2731–2740. 10.1145/2207676.2208670

Hagemoen, C. (2018). Hunting Vancouver's forgotten sidewalk prisms. *Scout Magazine*. Retrieved 23 June 2020 from https://scoutmagazine.ca/2018/04/30/hunting-vancouvers-forgotten-sidewalk-prisms/

Jewell, N. (2018). Translucent 'hugging' towers could help clean Hong Kong's air pollution. *In Habitat*. Retrieved 22 June from https://inhabitat.com/translucent-hugging-towers-could-help-clean-hong-kongs-air-pollution/

Kaiying, C. L., Plewe, D. A., & Röcker, C. (2015). The ambience of ambient intelligence: An Asian approach to ambient systems? *Procedia Manufacturing*, *3*, 2155–2161. doi:10.1016/j.promfg.2015.07.355

Lehtovuori, P. (2010). *Experience and conflict: The production of urban space*. Ashgate Publishing.

Lewis, J. (2009, Mar. 15). New glass city. *The New York Times*.

Materials. (2019). What exactly is a polycarbonate translucent façade? *ArchDaily*. Retrieved 12 June 2020 from https://www.archdaily.com/922735/what-is-exactly-a-polycarbonate-translucent-facade

McDonald, D. W., Gokhman, S., & Zachry, M. (2012). Building for social translucence: A domain analysis and prototype system. *CSCW '12: Proceedings of the ACM 2012 conference on Computer Supported Cooperative Work*, 637–646. 10.1145/2145204.2145301

McLeod, K., Spachos, P., & Plataniotis, K. (2020). An experimental framework for wellness assessment through the Internet of Things. *IEEE Internet Computing*, *24*(2), 8–17. Advance online publication. doi:10.1109/MIC.2020.2972817

Morgan, A. D. (2011). Journeys into transformation: Travel to an "other" place as a vehicle for transformative learning. *Journal of Transformative Education*, *8*(4), 246–268. doi:10.1177/1541344611421491

Nisenbaum, A. (2019). What's holding smart cities back? *Scientific American*, *18*(March). Retrieved June 10, 2020, from https://blogs.scientificamerican.com/observations/whats-holding-smart-cities-back/

Panetta, K. (2019). Gartner top 10 strategic technology trends for 2020. *Smarter With Gartner*. Retrieved 7 March 2020 from https://www.gartner.com/smarterwithgartner/gartner-top-10-strategic-technology-trends-for-2020/

Prestigiacomo, R., Hadgraft, R., Hunter, J., & Locker, L. (2020). *Learning-centered translucence: An approach to understand how teachers talk about classroom data. In LAK'20*. ACM.

Ringuette, J. (2007). Walking over history: Victoria's historic sidewalk prisms. Photos by N. Ringuette. In *Hidden in plain sight: Heritage and art features on Victoria street and sidewalks*. IslandNet. Retrieved 22 June 2020 from https://web.archive.org/web/20080509074924/http://www.islandnet.com/~jar/streetscapes/topics/prisms.htm

Smalley Bowen, T. (2007). Glass: Transparent, translucent, and ironic. *Architectural Record, 15*(October). Retrieved June 12, 2020, from https://www.architecturalrecord.com/articles/4011-glass-transparent-translucent-and-ironic

Stein, M., Draper, P., & Hellyer, R. (2019). City of glass: Recent advancements in glass structures in New York City. *Structural Engineering International, 29*(1), 101–111. doi:10.1080/10168664.2018.1519365

Steiner, D. (2011). Civic pride. *The Canadian Architect, 56*(10), 12–16.

Streitz, N., Charitos, D., Kaptein, M., & Böhlen, M. (2019). Grand challenges for ambient intelligence and implications for design contexts and smart societies. *Journal of Ambient Intelligence and Smart Environments, 11*(1), 87–107. doi:10.3233/AIS-180507

Times Hudson Valley. (2018). Translucent images at Garrison. *Times Hudson Valley Media.* Retrieved 22 June 2020 from http://timeshudsonvalley.com/stories/translucent-images-at-garrison,497?

Wang, L. (2019). An auditorium uses translucent ETFE panels for a surreal look. *InHabitat.* Retrieved 22 June 2020 from https://inhabitat.com/an-auditorium-uses-translucent-etfe-panels-for-a-surreal-look/

Weiser, M., & Seely Brown, J. (1995). *Designing calm technology.* Palo Alto, CA: Xerox PARC. Retrieved 9 June 2020 from https://people.csail.mit.edu/rudolph/Teaching/weiser.pdf

Wong, M. (2011). *Seattle prism light reconnaissance study.* Institute for Public Service, Seattle University.

Yin, R. K. (2018). *Case study research and applications: Design and methods* (6th ed.). Sage.

Zaiontz, C. (2019). *Real Statistics using Excel.* www.real-statistics.com

ADDITIONAL READING

CBC News. (2018). Want to build the city of the future? Here's your chance. Retrieved 24 May 2020 from https://www.cbc.ca/news/canada/british-columbia/smart-cities-vancouver-surrey-1.4560685

Celadyn, M. (2018). Environmental activation of inner space components sustainable interior design. *Sustainability, 10*(6), 1945. Advance online publication. doi:10.3390u10061945

Crowston, K., Saltz, J. S., Rezgui, A., Hegde, Y., & You, S. (2019). *MIDST: A system to support stigmergic coordination in data-science teams. CSCW'19 Companion*. ACM. doi:10.1145/3311957.3359509

Isberto, M. (2019). Ambient computing – the smart world. *Technology News*. Colocation America. Retrieved 17 April 2020 from https://www.colocationamerica.com/blog/ambient-computing-is-the-future-of-technology

McDonald, D. W., Beschastnikh, I., Kriplean, T., Borning, A., & Zachry, M. (2009). System design for social translucence in socially mediating technologies. Workshop on "Social mediating technologies: Developing the research agenda" at the ACM SIGCHI Conference on Human Factors in Computing Systems (CHI'09).

Naughton, J. (2020). For non-intrusive tracking of Covid-19, smartphones have to be smarter. *The Guardian*. Retrieved 29 April from https://www.theguardian.com/commentisfree/2020/apr/11/for-non-intrusive-tracking-of-covid-19-smartphones-have-to-be-smarter

Stevens, G. (2020). Computers without boundaries: A deep dive into ambient computing. *StreetFight Magazine*. Retrieved 16 March 2020 from https://streetfightmag.com/2020/03/16/computers-without-boundaries-a-deep-dive-into-ambient-computing/#.XnBVu1KZP6o

Whitaker, B. (2020). The computer algorithm that was among the first to detect the coronavirus outbreak. *CBS News, 60 Minutes*. Retrieved 28 April 2020 from https://www.cbsnews.com/news/coronavirus-outbreak-computer-algorithm-artificial-intelligence/

KEY TERMS AND DEFINITIONS

Ambient: In the context of smart cities, the ambient describes the increasing presence of more aware people together with awareness-enabled technologies in support of interactivity, influencing human communication and activities, human-machine interactivities, and machine-machine interactivities and in turn affecting behavior, experience, information capabilities, economies, literacies, and indeed everything.

Ambient Visibilities/Invisibilities: Ambient visibilities/invisibilities refer to urban elements accommodating more aware people and awareness-enabled technologies encompassing the physical, the digital, and the less tangible.

Awareness: Awareness is conceptualized in this work as the quality of being aware as it applies to people on the one hand, to the enabling of technologies on the other, and to the enhancing of awareness through people interacting with awareness-enabled and enabling technologies.

Invisibilities: Invisibilities refer to urban elements that are rendered invisible or that may present as intangibles.

Smart Cities: Smart cities are urban areas and regions characterized by more aware and engaged people, interacting with each other and aided by the use of more awareness enhancing technologies for a wide range of purposes from communication to design to atmospheres to mobility to livability, to governance, to data generation and use for learning and informed decision-making.

Translucence: Translucence refers to greater ease in seeing through situations, materials, the digital, and the like, in relation to smart cities and regions.

Visible Infrastructures: Visible infrastructures refer to physical and other visible urban elements.

Section 3
Theory and Methods

Chapter 5
InVisibilities:
Developing Spaces for Theory in Smart Cities

ABSTRACT

The purpose of this chapter is to explore urban visibilities and invisibilities, being attentive to the physical, digital, and ambient, giving rise to the need for further developing and extending spaces for theory. As such, this chapter seeks to shed light on the importance of elements pertaining to the visible, the invisible, and the ambient in urban environments informing theory for smart, responsive, and future cities. The research literature for urban theory in relation to urban visibilities and invisibilities is explored in this chapter in the context of smart and responsive cities. Using an exploratory case study approach, this chapter makes a contribution to 1) the research literature for urban theory in relation to smart and responsive cities, 2) the evolving of urban theory for smarter cities and regions, and 3) the evolving of theory through frameworks in support of the interweaving of spaces for invisibilities and visibilities as InVisibilities in a more aware environment, as in, ambient.

1. INTRODUCTION

In light of emerging relationships between visibilities and invisibilities (as presented in Chapters 1 and 2) and the associated dimensions of transparency, translucence, and the ambient (as presented in Chapters 3 and 4) in smart

DOI: 10.4018/978-1-7998-3850-0.ch005

cities and regions, this chapter explores the potential for developing spaces for theory in accommodating the ambient as a key element. That urban theory is said to be underdeveloped (Batty, 2013; Roy, 2009) and "in want of a good theoretical base" (Harrison and Donnelly (2011), gives rise to the need for urban theory development in relation to smart cities and more particularly, in support of the dimensions of visibilities and invisibilities. Indeed, while pointing to the many definitions of smart cities (Praharaj and Han, 2019; UN Habitat III, 2015) researchers lament the absence of a unique (Neirotti, De Marco, Cagliano, Mangano, and Scorrano, 2004) or agreed upon definition (Kuru and Ansell, 2020), contributing to "fuzziness" that "hinders our understanding" (Lara, Da Costa, Furlani, and Yigitcanlar, 2016) while Ramaprasad, Sánchez-Ortiz, and Syn (2017) claim the term has "more than thirty-six definitions" giving rise to the need for a unified definition which they provide through proposing an ontology for the domain. Guiding the investigation in this chapter is the use of an exploratory case study along with an explanatory correlational design where data are collected through the use of multiple methods including survey and in-depth interviews. This work is significant in that it argues for understanding smart cities through the lens of visibilities and invisibilities. In terms of contributions, this chapter a) extends the research literature for invisibilities and visibilities in smart cities from a theoretical perspective; b) evolves theory for smart cities and regions; and c) formulates a conceptual framework for InVisibilities and the ambient in developing new spaces for theory in smart cities. This chapter also contributes to the development potential for ambient theory, offering a novel perspective that accommodates characteristics such adaptability, awareness, and complexity, to name a few, in possibly extending existing urban theory. As such, the objective of this chapter is described below.

Objective: The objective of this chapter is to explore theories of cities, the urban, and smart cities pertaining to visibilities and invisibilities as InVisibilities, in further conceptualizing ambient theory for smart cities. As such, the key research question posed is – *Why does ambient theory matter for cities, the urban, and smart cities?*

2. BACKGROUND AND OVERVIEW

Roy (2009) articulated the need for 'new geographies' in terms of "imagination and epistemology in the production of urban and regional theory" to accommodate 21st century cities. Batty, Hudson-Smith, Hugel, and Roumpani (2015) claim that, "no good theories of the smart city in the short- or long-term exist." Brenner and Schmid (2015) advance the need for "a new epistemology of the urban" enabling a moving beyond "conceptions of the urban as a fixed, bounded and universally generalizable settlement type." The combination of theory, experiment, and simulation are said by Schmitt (2015) to be needed "to better understand the city." Wise and Shaffer (2015) note the importance of theory in assisting to determine "what variables a researcher should attend to" as well as "how to interpret a multitude of micro-results and make them actionable." Farina, Kotsopoulos, and Casalegno (2018) highlight the need "for the integration of a digital, immaterial dimension introducing new variables and expertise" in "hybrid connected spaces" between the physical and digital. With such variables in mind, it is worth considering the concerns posed by Stephanidis et al. (2019) where human activities in the form of "location, postures, emotions, habits, intentions, culture and thoughts" become "candidate input commands to a variety of visible and invisible technological artifacts embedded in the environment."

2.1 Definitions

Definitions are presented here for key terms used based on the focus of the literature reviewed and the research conducted in this chapter. Additional terms with definitions, as articulated by this work, are included in the *Key Terms and Definitions* section at the end of this chapter.

Ambient Theory:. McKenna (2021) advances ambient theory for smart cities consisting of the three elements of "awareness in relation to people and technologies; awareness-based spaces that are adaptive, dynamic, emergent, interactive, and pervasive; and meaningfully involving people in action."

InVisibilities: The InVisibilities concept highlights the interweaving, interconnectedness, and interdependencies of invisibilities and visibilities in smart cities and regions.

Urban Theory: Urban theory is extended in this work so as to conceptualize smart cities and regions as encompassing understandings of both people and technologies as smart, that are augmented, enriched and enhanced by each other.

2.2 Overview

This chapter provides perspectives in Section 3 on theory pertaining to invisibilities and visibilities, along with the ambient in smart cities through a review of the research and practice literature. A conceptual framework for invisibilities and visibilities as InVisibilities and the ambient is then developed and operationalized for use in the exploration conducted in Section 4. Section 4 also includes a description of the methodology used in the exploration followed by the results and a discussion of findings. Section 5 describes opportunities associated with future directions for research and practice. Section 6 concludes with chapter coverage highlights, the major contributions, limitations, and key insights and ideas. Several exercise questions are also included for the reader in support of further exploration.

The primary audiences for this chapter include students, educators, researchers, policy makers, businesses, community members, and urban practitioners and policymakers concerned with spaces for theory that support working with visibilities and invisibilities associated with urban design, urban experiences and interactions, and the quality of everyday life in the context of smart cities, regions, and future cities.

3. PERSPECTIVES ON INVISIBILITIES AND THE AMBIENT IN SMART CITIES

In the context of smart cities, this chapter provides a review of the research literature for emerging spaces for theory, accommodating invisibilities and visibilities as InVisibilities in relation to the physical, digital, and intangible and associated elements pertaining to urban ambiences and atmospheres; theory of cities; and change, complexity, and constraints.

3.1 Invisibilities and Visibilities and Theory in Smart Cities

Invisibilities and visibilities are considered to be highly interwoven in this work, as InVisibilities, in the context of smart cities and this chapter provides a review of the research literature for emerging spaces for theory.

Ambiences and Atmospheres

Kaiying, Plewe, and Röcker (2015) explore ambient intelligence (AmI) in terms of ambience from an Asian perspective focusing on "how aesthetic intelligence effects local ambient technologies." Demers and Potvin (2016) employ a physical ambiances laboratory as an "adaptable structure in an outdoor environment" in pursuing "the aims to connect the theory of ambiances with the actual complexity of experiencing on a site." Where Friberg (2014) spoke of atmosphere "as an aspect of perception", Christiansen, Laursen, and Hvejsel (2017) make a connection between ambiance and atmosphere where "the delicate notion of ambiance" can be "understood as a means for describing the atmosphere of a given space." Bille (2018) describes atmosphere in terms of "that through which the world appears." De Matteis, Bille, Griffero, and Jelić (2019) introduce a series of articles addressing various aspects of a theory of atmospheres from the perspective of phenomenographies. Phenomenography is described as "the differed representation of a lived situation, actual or envisioned, relying on one or more intersecting expressive media – including written narrative – encompassing the spatial, corporeal and affective spheres."

Theory of Cities

Augoyar (2002) advanced a theory integrating architectural and urban ambient environments, as in, urban ambience. According to Harrison and Donnelly (2011) "a key enabler for new theories of cities" is that of instrumentation in that "the ability to observe many individual behaviours in 'real time'," and "the way that people use the city" contributes to "making the invisible visible" requiring a reconsideration of "the emerging roles of information and technology" that "may enable new approaches to theories of cities." Yet, Harrison and Donnelly (2011) acknowledge the pervasiveness of information systems in urban environments enabling a "change" in terms of "opportunities to capture information" yielding real time access at the individual level contributing to the need "to address serious issues of privacy and protection of Personal Information." Van Nes (2014) uses space syntax to contribute to theory on built environments focusing on "relationships between space and society." Connolly, Keil and Harris Ali (2020) build upon urban theory by providing "an initial attempt to theorise the relationship between processes of extended urbanisation and infectious disease" highlighting concerns with population density.

Change, Complexity, and Constraints

Ibrahim, El-Zaart, and Adams (2018) employ the theory of change (ToC) in developing a transformation roadmap for the implementation of smart cities, taking readiness into consideration. Ibrahim (2019) works toward further validation of the ToC framework for use in smart sustainable cities (SSCs). The reach of smart cities across the complex intermingling of the physical layer, the services layer, and the digital layer is described by Finger (2016). Stevens and Salmon (2017) identify the need for understanding "cities as complex sociotechnical systems" in order "to optimise the interactions between people, technology, and environments" where constraints are identified and leveraged. Stokes (2006) speaks of constraints in relation to creativity and the value of constraints as stimulus, referring to the thinking of Calvino and his work (1974) in terms of a discussion between Kublai Khan and Marco Polo in *Invisible Cities* about an "endless and formless" empire structured by 18 conversations using patterns and repetition to describe the network of 55 imperial cities. Shutters (2018) points to "the immense complexity of cities" as a theme in a special issue focusing on the science of cities. Grieten, Lambrechts, Bouwen, Huybrechts, and Cooperrider (2018) probe appreciative inquiry, explained as an approach to inquiry based on "theory that anticipates the future" and "opens up possibilities or new forms of action" as in, "a theory of possibility" in support of "non-deficit positive change."

Matheus and Janssen (2019) advanced the use of window theory, highlighted in Chapter 3 of this book, in relation to transparency as an integrative theory. It may be that window theory could serve a broader role as a unifying theory in smart cities more generally.

Figure 1 provides an overview of theories for smart cities emerging from the research literature pertaining to visibilities and invisibilities and encompasses ambiences, atmospheres, change, cities, constraints, and the social with the potential for unity afforded possibly by window theory.

Where interactions in smart city environments involve artificial intelligence (AI), it will be important to consider autonomy, taking people into consideration and the Human-Centered Artificial Intelligence (HCAI) framework advanced by Shneiderman (2020), moving from 1-dimensional to 2-dimensional thinking.

Figure 1. Overview of theory for smart cities

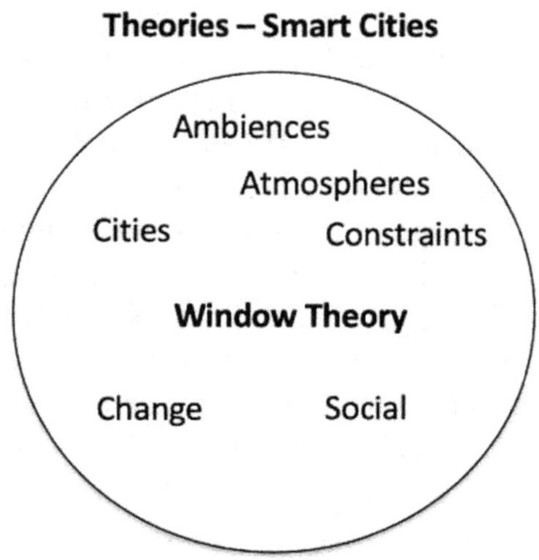

3.2 Summary

Table 1 provides a summary by year and author for theory in the context of visibilities and invisibilities in smart cities and regions covering the timespan of 2002 to 2020.

Over this time period, theory appears in relation to ambiences pertaining to ambient intelligence (AmI) (Kaiying et al., 2015); complexity (Demers and Potvin, 2016); and atmospheres (Christiansen et al., 2017). Theory appears in relation to atmospheres pertaining to perception (Friberg, 2014); ambiance (Christiansen et al., 2017); "that through which the world appears" (Bille, 2018); and phenomenography (De Matteis et al., 2019). Theories of cities emerge in relation to urban ambient environments (Augoyar, 2002); constraints in relation to creativity (Stokes, 2006); changes enabled by instrumentation (Harrison and Donnelly 2011); and density in relation to infectious disease (Connolly et al., 2020). Theory of change emerges in relation to readiness (Ibrahim et al., 2018) with further validation by Ibrahim (2019) while appreciative inquiry (Grieten et al., 2018) takes an approach to change involving possibility. Complexity theory is addressed by Finger (2016) in relation to infrastructures; by Stevens and Salmon (2017) as sociotechnical systems; and by Shutters (2018) in terms of the sheer immensity of the complexities of cities.

Table 1. Overview of key elements characterizing InVisibilities and theory in smart cities

Author(s)	Year	Ambiences	Atmospheres	Cities	Change	Complexity
Augoyar	2002	Urban		Ambience		
Stokes	2006			Constraints		
Harrison & Donnelly	2011			Change		
Friberg	2014		Perception			
Kaiying et al.	2015	AmI				
Demers & Potvin	2016	Complexity				
Finger	2016					Infrastructures
Christiansen et al.	2017	Atmospheres	Ambiances			
Stevens & Salmon	2017					Sociotechnical
Bille	2018		World			
Grieten et al.	2018			Appreciative Inquiry		
Ibrahim et al.	2018				Readiness	
Shutters	2018					Immensity
De Matteis et al.	2019		Phenomenography			
Ibrahim	2019				Validate	
Connolly et al.	2020			Density		
Shneiderman	2020	HCAI – reliable, safe and trustworthy				

Also of note is the presence of complexity across various theories and also, the notion of constraints.

Relevant to all theories is perhaps the work of Shneiderman (2020) addressing the importance of human-centered artificial intelligence (HCAI) focusing on reliable, safe, and trustworthy.

Across this array of elements characterizing InVisibilities and theory in smart cities, it is worth noting that ambience, change, and appreciative inquiry appear for more than one theory. Also of note is the presence of AmI and complexity.

The terms in Table 1 inform and guide the selection of variables and proxies that are used in exploring relationships for visibilities and invisibilities and the ambient in support of the evolving of theory for smart cities in this chapter.

3.3 InVisibilities and the Ambient: New Spaces for Theory in Smart Cities

Based on the review of the research literature presented in Section 3.1 of this chapter, a conceptual framework for invisibilities and visibilities as InVisibilities and the ambient is advanced to accommodate spaces for theory as depicted in Figure 2.

The interweaving of visibilities and invisibilities in smart cities focusing on characteristics of the ambient such as adaptability, awareness, augmentability, attentiveness, attuning, and autonomy contribute to evolving spaces for urban theory, ambient theory, and the notion of ambient InVisibilities.

Figure 2. Framework for InVisibilities and the ambient: new spaces for theory in smart cities

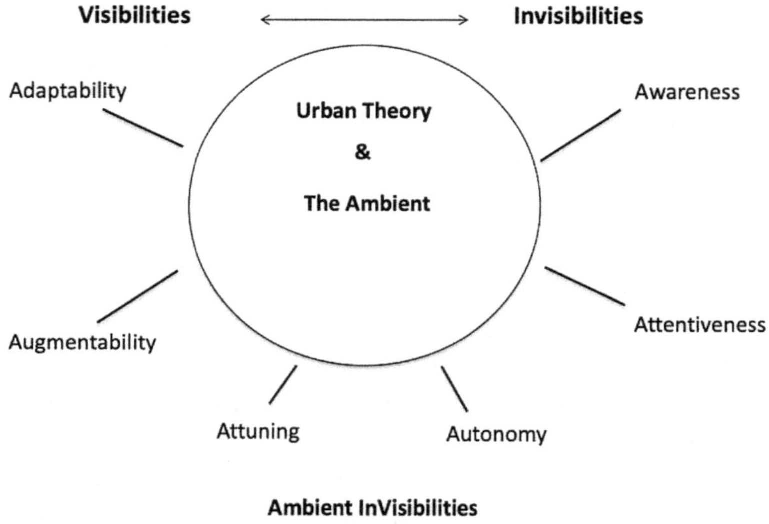

This framework contributes further to guidance for exploration of the research question for this chapter presented in Section 1 and reformulated here as a proposition, as stated below.

P1: Ambient theory matters for cities, the urban, and smart cities in accommodating explorations and implementations of emergent and evolving spaces for an interweaving of visibilities and invisibilities as InVisibilities in relation to the ambient.

How evolving spaces for theory inform visibilities and invisibilities is explored in this chapter in the context of smart cities. As such, under exploration is awareness and other elements characterizing the ambient, based on people and their experiences. Ambient spaces for theory are understood to be more dynamic, in-the-moment, and adaptive, enabling new potentials for urbanizing to emerge. As such, spaces for theory in relation to InVisibilities as ambient are discussed in the context of contemporary and future smart cities that are posing unprecedented challenges and opportunities. While McKenna (2019) refers to theory as ambient, ambient theory was neither proposed nor advanced until more recently (McKenna, 2021) in relation to awareness encompassing people and technologies, awareness-based spaces, and engaging people in meaningful action. This work further supports and advances the need for ambient theory for smart cities based on the conceptualization in Figure 2.

4. METHODOLOGY FOR EXPLORING AMBIENT IN/VISIBILITIES IN SMART CITIES

Based on the perspectives emerging from the review of the research literature in this chapter, smart city theory and the ambient are explored here in relation to visibilities and invisibilities as InVisibilities. The proxies of *information and communication technologies (ICTs)* and *awareness* are used in this exploration for *InVisibilities* and *the ambient*, respectively.

4.1 Research Design for Exploring Ambient InVisibilities in Smart Cities

An exploratory case study approach was used as recommended by Yin (2018) for the study of contemporary phenomena, in this instance, InVisibilities and the ambient in smart cities and regions. The research design in this chapter also incorporates an explanatory correlational design, said by Yin (2018) to accommodate "complementarity of case study and statistical research." A website was used to describe the study and during sign up, demographic data were collected including elements such as age range, location (city), and people could self-identify in one or more categories (e.g., community member, city official, researcher, educator, student, business, etc.). Upon sign up, people were invited to complete a survey about their experience of their city as a smart city. An invitation was also provided to participate in an in-depth interview about smart cities, online or in person, depending upon location. The study attracted participation from across Canada, extending to cities in the United States, Ireland, and Israel, to name a few. The survey instrument, containing a mix of closed and open-ended questions, and the interview protocol were pretested prior to use in the study. To enable responses from a broad spectrum of people, heterogeneity sampling, a type of purposive sampling was used whereby an invitation to participate in the study was posted to webspaces with the potential to attract a wide response. Data analysis included the use of content analysis for qualitative data and descriptive statistics for quantitative data. The Real Statistics Software Resource Pack Add In for Microsoft Excel (Zaiontz, 2019) was used, consisting as it does of a wide range of tools for statistical analysis, such as the correlation feature for calculation of the Spearman correlation for the ordinal data generated by assessments from survey respondents. The survey instrument was designed to enable people to assess their experience of the city and region where they live in terms of a wide variety of elements including ICTs and awareness. Data were also systematically collected in parallel with this study through group and individual discussions with a wide range of people across many sectors in small to medium to large-sized cities in Canada (e.g., greater Victoria, Vancouver, and Toronto). Overall, the analysis of data consisted of n=78 including 41% females and 59% males for people ranging in age from their 20s to their 70s.

4.2 Exploring Spaces for InVisibilities and the Ambient in Smart Cities

When asked to assess to what extent elements such as information and communication technologies (ICTs) and awareness contribute to the making of smart cities on a scale of 1 (Not at all) to 7 (absolutely), as indicated in Table 2, 50% responded at position 5 on the 7-point Likert-type scale for I*CTs*, and 50% at the upper end of the scale at position 7. For *awareness*, 25% responded at position 5, 25% at position 6, and 50% at position 7.

In determining the correlation between *InVisibilities* using *ICTs* as a proxy where such technologies may be visible in some cases and invisible in others and *the ambient* using the proxy of *awareness*, the correlation feature of the

Table 2. Assessments for InVisibilities (ICTs) and the Ambient (awareness) in smart cities

Items	Assessments	Correlation (Spearman)
InVisibilities (ICTs)	50% (5); 50% (7)	-.23
The Ambient (awareness)	25% (5); 25% (6); 50% (7)	

Real Statistics Software Add In (Zaiontz, 2019) for Microsoft Excel is used showing the Spearman correlation coefficient for ordinal data result of -.23. Creswell (2018) advises that a correlation in the .20 to .35 range is considered "only a slight relationship" and in this case, the direction of the relationship is a negative, inverse one.

Looking at the scatterplot for this relationship depicted in Figure 3, as the perception of ICTs (InVisibilities) decreases or levels out, the perception of awareness (the ambient) would seem to increase suggesting the presence possibly of more complex factors at play, requiring further and deeper probing going forward, with the use of other variables and perhaps other variable combinations.

Qualitatively, through discussion and reflection, a community member, educator, and person in business highlighted the importance of "the technology" and "being smart on how we use it" adding that the smart cities concept is "a great idea because it's both theoretical and practical." Another educator

Figure 3. Scatterplot for InVisibilities (ICTs) and the Ambient (awareness) in smart cities

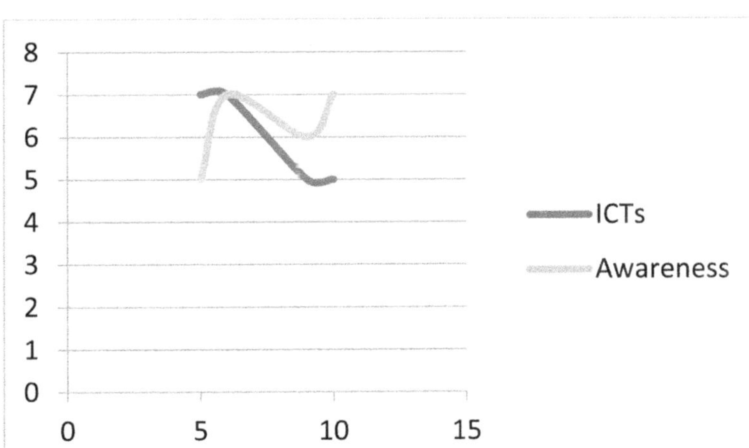

and person in business pointed to the "both together" potential, inclusive of "human values and needs" with "IoT, Big Data, and other tech systems to make the city a better place." The use of real-time streaming was noted by a city councilor as a way of appropriating technology for eTownHall meetings enabling people to attend virtually while providing comments and feedback and generating additional interest and engagement in such events through social media. In terms of creating atmosphere and a sense of vibrancy in the city, an educator referred to "human activity in every corner" through everyday life and urban events in the city.

In further support of ambient spaces for theory, collaborative art and technology initiatives in public places were described in the form of a musical railing parkade project with a community leader noting that, "there is a lot of theory around what makes vibrant spaces." Also of note were discussions of an urban display to capture and display cycling data on a visual screen in-the-moment for a multi-use path. A city councilor described BikeMaps.org as "a great example" of how "its' all connected, and the tool of technology is augmenting the information that we have and allows us to have a safer more vibrant city."

4.3 Discussion and Interpretation of Findings

Findings in this chapter serve to challenge the notion of ambient spaces for theory in coming to evolving understandings of the interweaving of visibilities and invisibilities in smart cities in light of the presence of an inverse "slight relationship" (Creswell, 2018) between *InVisibilities (ICTs)* and *the ambient (awareness)*. However, it may be that this inverse, slight relationship highlights the complexities and uncertainties of smart cities. For example, in terms of ambiances and complexities, the eTownHall event described by the city councilor serves to highlight how technology enables people, as well as city hall meetings and processes, to become more visible, contributing to increased awareness through social media and other ICT-based approaches. The pulse and vibrancy of the city emerge in relation to understandings of urban atmospheres as revealed by an educator describing the physical and digital and a city councilor describing interactions around art, culture, people and technologies. Complexities emerge in relation to appropriating the concept of smart cities, being smart about smart technologies, and leveraging information while maintaining privacy, to name a few. If, as Panetta (2019) predicts, the "ambient experience" will encompass one of Gartner's "top

10 strategic technology trends" as the "multiexperience" then the need to understand the ambient becomes critical as does the need for ambient theory.

InVisibilities have the potential to contribute further to the awareness-based ambient implementation explanatory model for smart cities (McKenna, 2020) in relation to one or more of the variable sets such as the *awareness and complexity* dependent variable (DV). *InVisibilities* and *the ambient* may serve as a response to the call by Farina et al. (2018) for a design variable pertaining to the "digital, immaterial dimension." Further, *InVisibilities* and *the ambient* may have the potential to be key influencers for new spaces for theory in support of opportunities associated with ambiences, atmospheres, change, and complexities.

5. EMERGING RESEARCH AND OPPORTUNITIES

Future and emerging trends are identified in this chapter for practice and research in Sections 5.1 and 5.2 pertaining to InVisibilities and the ambient in smart cities and regions.

5.1 Practice

Practically speaking, this chapter contributes to the rethinking of urban elements that could serve as variables to explore that would inform theory in order to make spaces for addressing challenges associated with the interweaving of the visible and invisible in smart cities and to the surfacing of relationships among variables for understandings of the ambient.

Ambient InVisibilities

This chapter provides exposure to explorations of urban spaces in addressing smart city visibilities and invisibilities in relation to ICTs and awareness. Crucially, this work invites cross-sector participation in the development of initiatives in support of urban problem-solving that are more adaptive and responsive to urban challenges and opportunities, focusing on elements pertaining to atmospheres, change, complexities and the like, characterizing ambient and smart environments.

5.2 Research

This chapter advances the potential for future research opportunities and emerging discourse spaces for visibilities and invisibilities in relation to urban theory and the ambient in the context of smart cities.

Ambient Spaces for Theory

The viability of the conceptual framework for InVisibilities and the ambient in support of new spaces for theory in smart cities operationalized for use in this chapter demonstrates early-stage potentials while offering and opening up new understandings for evolving urban spaces for testing and practice.

Ambient InVisibilities

Conceptualization of spaces for real-time, dynamic, and adaptive types of thinking for understanding the invisible and visible emerge in this chapter opening the way for possibly novel roles and collaborations for citizens, businesses, researchers, students, and governments. As such, areas for exploration potential include:

1. *Ambiences* fostered through urban initiatives focusing on awareness in public spaces with dynamic and interactive visualizations and augmentations
2. *Atmospheres* fostered through attentiveness to a sense of awareness of the city
3. *Complexities* understood through extending meetings, services, and communications with technology-based options for interactions

Ambient Theory

This work contributes to the notion of ambient theory as a potentially novel addition to existing theories for urban research, amenable to adaptive frameworks for smart, responsive, and future cities. Ambient theory would enable possibly more adaptive, dynamic, interactive, and evolving models in response to increasing complexities.

6. CONCLUSION

This chapter opens spaces for consideration of ICTs (information and communication technologies) and the ambient as dimensions of visibilities and invisibilities for smart cities in relation to what is emerging from a range of disciplinary perspectives involving the innovating of spaces for urban theory. Ambiances, atmospheres, and change emerge as key theories for navigating invisibilities and visibilities in smart environments. Spaces for these theories will be enabled through fostering increased education and adaptive thinking. The main contributions of this chapter include: a) articulation and operationalization of a conceptual framework for InVisibilities and the ambient in developing new spaces for theory in smart cities; b) further development of the research literature for visibilities and invisibilities in smart environments by the extending of dimensions to include *InVisibilities* and *the ambient* represented in this work by the proxies of *ICTs* and *awareness*; and c) highlighting the elements of ambiances, atmospheres, and complexities in fostering smarter, more responsive, cities and regions. As such, this work contributes to the notion of ambient theory as a framework for understanding smart cities, taking aware people and aware technologies into consideration.

Limitations of this chapter pertain to the small sample size of the study upon which this exploration is based. Additionally, limitations of this chapter related to the use of *ICTs* and *awareness* as proxies for the exploration of *InVisibilities* and *the ambient* could be mitigated going forward by exploring the latter more directly or by using other variables in future studies. And limitations related to literature review domain coverage are mitigated by the interdisciplinary potential to extend this review to other areas going forward.

This chapter continues to give form, detail, and definition to the exploration of dimensions of visibilities and invisibilities in smart cities and regions. The analysis of findings identifies a correlation of -.23 between ICTs and the ambient, constituting what Creswell (2018) describes as a "only a slight relationship" and an inverse one at that, which is nevertheless relevant to the study of emergent and evolving spaces for theory in smart cities and potentials for ambient theory.

Insights of particular interest to the reader include the following:

1. The emerging range of dimensions for invisibilities and visibilities in the form of *ICTs* as *InVisibilities* and *awareness* as *the ambient* in smart cities

2. The presence of a "slight" inverse relationship emerged between *InVisibilities* (e.g., *ICTs)* and *the ambient* (e.g., *awareness)* when correlated in a smart cities context
3. The interplay of ambience, atmospheres, and complexities associated with *InVisibilities* and *the ambient*

Ideas highlighted in this chapter pertain to the following:

1. The importance of people becoming more involved in understandings of *ICTs* and *the ambient* as dimensions of invisibilities and visibilities in smart cities
2. The potential for people to shed light on *ICTs* and *awareness* in urban spaces
3. Explorations of *ICTs* and *the ambient* as novel research spaces informing theory for smarter cities and regions

Exercises for the reader to consider, encouraging further exploration include the following:

1. Identify one or more examples of *InVisibilities* and *ICTs* in your city
2. Identify one or more examples of *the ambient* and *awareness* in your city
3. Describe why *ICTs* and *awareness* could be important for smart cities, based on your example(s)

If you would like to share your responses to the Exercises with the author, please complete the questions in this online space for Chapter 5 – InVisibilities and Theory

This work will be of interest to a broad audience including students, educators, researchers, artists, urban planners, urban practitioners, community members, and anyone concerned with emerging spaces for theory that take into consideration the complex urban challenges and opportunities posed by renderings of the invisible, visible, and the ambient in smart, responsive, and future cities.

REFERENCES

Augoyard, J.-F. (2002). Deficiencies in the research about architectural and urban ambient environment and proposition of new tools: The sound example. *Architectural and Urban Ambient Environment: First International Workshop.*

Batty, M. (2013). Big data, smart cities and city planning. *Dialogues in Human Geography, 3*(3), 274–279. doi:10.1177/2043820613513390 PMID:29472982

Batty, M., Hudson-Smith, A., Hugel, S., & Roumpani, F. (2015). Visualising data for smart cities. In *Handbook of Research on Social, Economic, and Environmental Sustainability in the Development of Smart Cities*. IGI Global. doi:10.4018/978-1-4666-8282-5.ch016

Bille, M. (2018). Review: Gernot Böhme, 2017, The aesthetics of atmospheres, Edited by Jean-Paul Thibaud. London, Routledge. Ambiances: International Journal of Sensory Environment, Architecture and Space, 1065.

Brenner, N., & Schmid, C. (2015). Towards a new epistemology of the urban? *City, 19*(2-3), 151–182. doi:10.1080/13604813.2015.1014712

Calvino, I. (1974). Invisible cities. New York: Harcourt. (Original publication 1972)

Christiansen, E. M., Laursen, L. H., & Hvejsel, M. F. (2017). Tectonic perspectives for urban ambiance? Towards a tectonic approach to urban design. *Ambiances: International Journal of Sensory Environment, Architecture and Space*, 886. Retrieved 21 April 2018 from https://journals.openedition.org/ambiances/886

Connolly, C., Keil, R., & Harris Ali, S. (2020). Extended urbanisation and the spatialities of infectious disease: Demographic change, infrastructure and governance. *Urban Studies (Edinburgh, Scotland)*, 1–19. doi:10.1177/0042098020910873

Creswell, J. W. (2018). *Educational Research: Planning, Conducting, and Evaluating Quantitative and Qualitative Research* (6th ed.). Pearson.

De Matteis, F., Bille, M., Griffero, T., & Jelić, A. (2019). Phenomenographies: Describing the plurality of atmospheric worlds. *Ambiances,* 5. Retrieved 25 April 2020 from https://journals.openedition.org/ambiances/2526

Demers, C., & Potvin, A. (2016). From history to architectural imagination: A physical ambiences laboratory to interpret past sensory experiences and speculate on future spaces. *Ambiances: International Journal of Sensory Environment, Architecture and Space*, 756.

Farina, C., Kotsopoulos, S. D., & Casalegno, F. (2018). Hybrid connected spaces: Mediating user activities in physical and digital spaces. In DAPI 2018 (LNCS 10921, pp. 35-55). Academic Press.

Finger, M. (2016). *Managing urban infrastructures*. Lausanne, Switzerland: MOOC, Fall.

Friberg, C. (2014). Performing everyday practices: Atmosphere and aesthetic education. *Ambiances: International Journal of Sensory Environment, Architecture and Space*, 464.

Grieten, S., Lambrechts, F., Bouwen, R., Huybrechts, J., Fry, R., & Cooperrider, D. (2018). Inquiring into appreciative inquiry: A conversation with David Cooperrider and Ronald Fry. *Journal of Management Inquiry*, *27*(1), 101–114. doi:10.1177/1056492616688087

Habitat, U. N., III. (2015). *Habitat III Issue Papers: 21 – Smart cities*. Retrieved 7 May 2020 from https://habitat3.org/wp-content/uploads/Habitat-III-Issue-Paper-21_Smart-Cities-2.0-2.pdf

Harrison, C., & Donnelly, I. A. (2011). A theory of smart cities. In *Proceedings of the 55th Annual Meeting of the International Society for the Systems Sciences*. University of Hull Business School.

Ibrahim, M. (2019). Developing smart sustainable cities: A validated transformation framework. *International Conference on Smart Applications, Communications and Networking (SmartNets)*, 1-5.

Ibrahim, M., El-Zaart, A., & Adams, C. (2018). Smart sustainable cities roadmap: Readiness for transformation towards urban sustainability. *Sustainable Cities and Society*, *37*, 530–540. doi:10.1016/j.scs.2017.10.008

Kaiying, C. L., Plewe, D. A., & Röcker, C. (2015). The ambience of ambient intelligence: An Asian approach to ambient systems? *Procedia Manufacturing*, *3*, 2155–2161. doi:10.1016/j.promfg.2015.07.355

Kuru, K., & Ansell, D. (2020). TCitySmartF: A Comprehensive Systematic Framework for Transforming Cities Into Smart Cities. *IEEE Access: Practical Innovations, Open Solutions*, *8*, 18615–18644. doi:10.1109/ACCESS.2020.2967777

Lara, A. P., Da Costa, E. M., Furlani, T. Z., & Yigitcanlar, T. (2016). Smartness that matters: Towards a comprehensive and human-centered characterisation of smart cities. *Journal of Open Innovation*, *2*(1), 8. doi:10.118640852-016-0034-z

Matheus, R., & Janssen, M. (2019). A systematic literature study to unravel transparency enabled by open government data: The window theory. *Public Performance & Management Review*. Advance online publication. doi:10.1080/15309576.2019.1691025

McKenna, H. P. (2019). Visibilities and invisibilities: Methodologies and theoretical spaces for ambient urbanities. In *Ambient Urbanities as the Intersection between the IoT and the IoP in Smart Cities* (pp. 192–222). IGI Global. doi:10.4018/978-1-5225-7882-6.ch007

McKenna, H. P. (2020). Awareness and smart city implementations: Sensing, sensors, and the IoT in the public sector. In J. R. Gil-Garcia, T. A. Pardo, & M. Gascó (Eds.), Beyond smart and connected governments: Sensors and the Internet of Things in the public sector. Springer.

McKenna, H. P. (2021). The importance of theory for understanding smart cities: Making a case for ambient theory. In N. Streitz & S. Konomi (Eds.), Lecture Notes in Computer Science: Vol. 12782. *Distributed, Ambient and Pervasive Interactions. HCII* (pp. 1–14). Springer. doi:10.1007/978-3-030-77015-0_4

Neirotti, P., De Marco, A., Cagliano, A. C., Mangano, G., & Scorrano, F. (2014). Current trends in Smart City initiatives: Some stylised facts. *Cities (London, England)*, *38*, 25–36. doi:10.1016/j.cities.2013.12.010

Panetta, K. (2019). Gartner top 10 strategic technology trends for 2020. *Smarter With Gartner.* Retrieved 7 March 2020 from https://www.gartner.com/smarterwithgartner/gartner-top-10-strategic-technology-trends-for-2020/

Praharaj, S., & Han, H. (2019). Cutting through the clutter of smart city definitions: A reading into the smart city perceptions in India. *City. Cultura e Scuola*, *18*, 100289. doi:10.1016/j.ccs.2019.05.005

Ramaprasad, A., Sánchez-Ortiz, A., & Syn, T. (2017). A unified definition of a smart city. In M. Janssen, K. Axelsson, O. Glassey, B. Klievink, R. Krimmer, I. Lindgren, P. Parycek, H. J. Scholl, & D. Trutnev (Eds.), *Electronic Government: 16th IFIP WP 8.5 International Conference, EGOV*. Springer. 10.1007/978-3-319-64677-0_2

Roy, A. (2009). The 21st century metropolis: New geographies of theory. *Regional Studies*, *43*(6), 819–830. doi:10.1080/00343400701809665

Schmitt, G. (2015). *Information Cities*. https://itunes.apple.com/us/book/information-cities/id970529491?mt=11

Shneiderman, B. (2020). Human-Centered Artificial Intelligence: Reliable, Safe & Trustworthy. *International Journal of Human-Computer Interaction, 36,* 6. doi:10.1080/10447318.2020.1741118

Shutters, S. T. (2018). Urban science: Putting the "smart" in smart cities. *Urban Science*, *2*(4), 94–97. doi:10.3390/urbansci2040094

Stephanidis, C., Salvendy, G., Antona, M., Chen, J. Y. C., Dong, J., Duffy, V. G., Fang, X., Fidopiastis, C., Fragomeni, G., Fu, L. P., Guo, Y., Harris, D., Ioannou, A., Jeong, K. K., Konomi, S., Krömker, H., Kurosu, M., Lewis, J. R., Marcus, A., ... Zhou, J. (2019). Seven HCI Grand Challenges. *International Journal of Human-Computer Interaction*, *35*(14), 1229–1269. doi:10.1080/10447318.2019.1619259

Stevens, N., & Salmon, P. (2017). Cities are complex systems – Let's start looking at them that way. *The Conversation*. Retrieved 4 May 2020 from https://theconversation.com/cities-are-complex-systems-lets-start-looking-at-them-that-way-72665

Stokes, P. B. (2006). *Creativity from constraints: The psychology of breakthrough*. Springer.

Van Nes, A. (2014). Space syntax in theory and practice. In Geodesign by Integrating Design and Geospatial Sciences. Springer. doi:10.1007/978-3-319-08299-8_15

Wise, A. F., & Shaffer, D. W. (2015). Why theory matters more than ever in the age of big data. *Journal of Learning Analytics*, *2*(2), 5–13. doi:10.18608/jla.2015.22.2

Yin, R. K. (2018). *Case study research and applications: Design and methods* (6th ed.). Sage.

Zaiontz, C. (2019). *Real Statistics using Excel*. www.real-statistics.com

ADDITIONAL READING

Dovey, K., & Pafka, E. (2016). Urban density matters – but what does it mean? *The Conversation*. Retrieved 29 April from https://theconversation.com/urban-density-matters-but-what-does-it-mean-58977

Flint, A. (2020). The coronavirus pandemic makes a case for megaregions. *CityLab*. Retrieved 30 April 2020 from https://www.citylab.com/equity/2020/04/coronavirus-reopen-economy-regional-planning-states-cities/610375/

Ho Chee Tat, T., & Loh Chee Ping, G. (2020). Innovating services and digital economy in Singapore. *Communications of the ACM*, *63*(4), 58–59. doi:10.1145/3378554

Mair, S. (2020). What will the world be like after coronavirus? Four possible futures. *The Conversation*. Retrieved 29 April from https://theconversation.com/what-will-the-world-be-like-after-coronavirus-four-possible-futures-134085

Nisenbaum, A. (2019). What's holding smart cities back? A tech-empowered urban renaissance is possible-if five challenges can be overcome. *Scientific American*, *18*(March). Retrieved May 4, 2020, from https://blogs.scientificamerican.com/observations/whats-holding-smart-cities-back/

Sadowski, J., & Pasquale, F. (2015). The spectrum of control: A social theory of the smart city. *First Monday*, *20*(7). Advance online publication. doi:10.5210/fm.v20i7.5903

Shenker, J. (2020). Cities after coronavirus: How covid-19 could radically alter urban life. *The Guardian*. Retrieved 29 April 2020 from https://www.theguardian.com/world/2020/mar/26/life-after-coronavirus-pandemic-change-world

KEY TERMS AND DEFINITIONS

Ambient: In the context of smart cities, the ambient describes the increasing presence of more aware people together with awareness-enabled technologies in support of interactivity, influencing human communication and activities, human-machine interactivities, and machine-machine interactivities and in turn affecting behavior, experience, information capabilities, economies, literacies, and indeed everything.

Ambient Theory: Ambient theory refers to an emerging new addition to existing theories for urban research that is more amenable to adaptive frameworks for smart, responsive, and future cities while enabling possibly more aware, dynamic, interactive, and evolving models in response to increasing complexities and uncertainties.

Ambient Visibilities/Invisibilities: Ambient visibilities/invisibilities refer to urban elements accommodating more aware people and awareness-enabled technologies encompassing the physical, the digital, and the less tangible.

Awareness: Awareness is conceptualized in this work as the quality of being aware as it applies to people on the one hand, to the enabling of technologies on the other, and to the enhancing of awareness through people interacting with awareness-enabled and enabling technologies.

Invisibilities: Invisibilities refer to urban elements that are rendered invisible or that may present as intangibles.

Smart Cities: Smart cities are urban areas and regions characterized by more aware and engaged people, interacting with each other and aided by the use of more awareness enhancing technologies for a wide range of purposes from communication to design to atmospheres to mobility to livability, to governance, to data generation and use for learning and informed decision-making.

Visibilities: Visibilities refer to physical and other visible urban elements and infrastructures.

Chapter 6
InVisibilities:
Methodological Spaces and Approaches in Smart Cities

ABSTRACT

The purpose of this chapter is to explore urban visibilities and invisibilities in terms of the physical, digital, less tangible, and ambient in relation to the need for new methodologies and approaches in understanding and designing for smart cities. This chapter seeks to shed light on the interweaving of elements in urban environments informing methodologies for smart cities. The research literature for urban methodologies and approaches in the context of visibilities and invisibilities is explored in this chapter for smart cities and regions. Using an exploratory case study approach combined with an explanatory correlational design, placemaking and attuning to urban spaces are investigated as proxies for InVisibilities and the ambient.

1. INTRODUCTION

Van Es and De Lange (2020) advance datawalking as a research method to address "invisibility, loss of context and access to data and its infrastructures" claiming that it "offers a much-needed anchoring of data as material and situated, and constitutive of everyday life." The study of smart cities has everything to do with daily life and Mehl and Connor (2012) provide a guide for studying many aspects of daily life, including several articles on the

DOI: 10.4018/978-1-7998-3850-0.ch006

importance of thinking in terms of "real-world" explorations. As such, this chapter seeks to identify and explore methodologies in support of research and practice in smart cities that take into consideration invisibilities and visibilities. Such methodologies draw upon the emerging dimensions of invisibilities and visibilities as presented in this work, together with the nature of relationships identified between these dimensions such as transparency, translucence, awareness, attuning, infrastructures, and information and communication technologies (ICTs), to name a few. This chapter is guided by the use of an exploratory case study along with an explanatory correlational design where data are collected through the use of multiple methods including survey and in-depth interviews. The significance of this chapter pertains to the approaches identified for understanding smart cities in relation to visibilities and invisibilities providing important insights as well as the potential for developing ambient inquiries. This chapter makes a contribution by a) extending the research literature for invisibilities and visibilities in smart cities from a methodological perspective; b) evolving approaches and methods for inquiries into visibilities and invisibilities in smart cities and regions; and c) formulating a conceptual framework for methodologies accommodating spaces for visibilities and invisibilities in smart cities. As such, the main objective of this chapter is described below.

Objective: The objective of this chapter is to explore methodologies for cities and urban regions pertaining to visibilities and invisibilities as InVisibilities in conceptualizing ambient approaches for understanding and working with smart environments. As such, the key research question posed is – *Why does ambient methodology matter for cities, the urban, and smart cities?*

2. BACKGROUND AND OVERVIEW

Crutzen and Hein (2009) identify several types of invisibility in the context of artificial intelligence (AI), as "mental, physical, and methodical" where "invisibility management could enable situated veiling and unveiling." Gavalas et al. (2017) identify methodologies associated with smart cities in the context of wireless connections and mobile computing based on the need for "novel methods of management" in support of sustainability. In the context of smart cities and smart citizens, Dembski et al. (2020) identify the need for approaches in urban planning "to cope with urban complexity" while accommodating "participatory and collaborative processes to empower citizens." Van Es and De Lange (2020) emphasize the centrality of *attentiveness* in the datawalking method "to data infrastructures making them visible for critical scrutiny and reflection."

2.1 Definitions

Definitions are presented here for key terms used, based on the focus of the literature reviewed and the research conducted in this chapter. Additional terms with definitions, as articulated by this work, are included in the *Key Terms and Definitions* Section at the end of this chapter.

Ambient Methodologies: This work advances ambient methodologies as approaches for the study and implementation of smart cities, taking aware people and aware technologies into consideration.

InVisibilities: The practice of InVisibilities as conceptualized in this work highlights approaches to working with the interwoven, interconnected, and interdependent nature of invisibilities and visibilities in smart cities and regions.

Urban Methodologies: Urban methodologies are extended in this work to conceptualize smart cities and regions as encompassing understandings of both people and technologies as smart, while accommodating spaces for enhancements, enrichments, and augmentations.

2.2 Overview

This chapter provides perspectives in Section 3 on methodologies pertaining to invisibilities and visibilities as InVisibilities, along with the ambient in smart cities through a review of the research and practice literature. A conceptual framework is then developed and operationalized for use in the exploration conducted in Section 4. Section 4 also includes a description of the methodology used in the exploration followed by the results and a discussion of findings. Section 5 describes opportunities associated with future directions for research and practice and Section 6 concludes with chapter coverage highlights, the major contributions, limitations, and key insights and ideas. Several exercise questions are also included for the reader in support of further exploration.

The primary audiences for this chapter include students, educators, researchers, policy makers, businesses, community members, and urban practitioners concerned with methods, approaches, and spaces for working with visibilities and invisibilities associated with urban design, urban planning, urban representation, and urban experiences, along with challenges and opportunities in the context of smart cities, responsive cities, and future cities.

3. PERSPECTIVES ON METHODOLOGY AND THE AMBIENT IN SMART CITIES

This chapter provides a review of the research literature for methodologies and approaches used in relation to the physical, digital, and the ambient, taking into consideration visibilities and invisibilities in the context of smart cities.

3.1 InVisibilities and Methodologies in Smart Cities

Invisibilities and visibilities are considered to be highly interwoven in this work, as InVisibilities, in the context of smart cities. This section provides a review of the research literature for a range of inquiry methods including appreciative inquiry, wonder, imagination, and transformation; datawalking, computation, artificial intelligence (AI) and machine learning (ML), and platforms; design thinking; experimentation, exploration, and experience; simulation, urban digital twins, and virtual reality (VR); space syntax, public spaces, and contexts; and self-relevant research.

Appreciative Inquiry, Wonder, Imagination, and Transformation

From an architectural perspective, Patel (2017) describes working with wonder as the frame ("an awe-inspiring moment at a distance" such as "a captivating view through a window"); the spectacle ("a circus" where "everything is designed to capture your imagination"), and the interruption ("the most subtle" which "is happened-upon" as in "the strange hidden in the familiar."). Grieten, Lambrechts, Bouwen, Huybrechts, and Cooperrider (2018) describe appreciative inquiry as both theory and method supporting an approach to change management in the context of organization development (OD) involving "a life-centric or strength-based (vs. deficit-based)" process. ICMA Press (2011) provides a guide for the application of appreciative inquiry in the context of urban environments, reframing problems as challenges and opportunities. Also concerned with potential and creativity, Dovey, Pafka, and Ristic (2018) explore "the city as a space of possibility" through questions such as "How can mapping extend the urban imagination and therefore the possibilities for urban transformation?" As a digital artist, Chevalier (2020) explores "virtual cities" and "the imagination of the city and its urban transformations" creating cities "between reality and simulation, forming part of a transformable space-time." Anderson (2019) advanced intuitive inquiry as a research method offering the potential for insights that contribute to transformation through "auspicious bewilderment."

Datawalking, Computation, AI/ML, Placemaking, and Platforms

It is worth noting that Gehl (2010) claimed "there is much more to walking than walking" adding that "it is also – and most particularly about – the city as meeting place" in support of "experiences and information" offering "a special form of communion between people who share public space as a platform and framework." Van Es and De Lange (2020) propose the notion of datawalking as a research method "to tackle some of the challenges that the computational turn has presented" such as, "*invisibility, decontextualization and* (lack of) *accessibility of data* and its infrastructures as objects of study" as in, "studying the datafied smart city." According to Van Es and De Lange (2020), datawalks "combine purposive physical walks through the (urban) landscape with being specifically attuned to observe and reflect on the variety of processes and infrastructures of datafication as situated in time and space." The Project for Public Spaces (PPS, 2007) describes placemaking as "both an overarching idea and a hands-on approach for improving a neighborhood, city, or region" that "inspires people to collectively reimagine and reinvent public spaces" in the form of a "collaborative process" whereby "we can shape our public realm" in ways that "facilitates creative patterns of use" attentive to "the physical, cultural, and social identities that define a place." As with datawalking, placemaking often involves physical walking through public spaces to collectively observe and reflect. In the context of artificial intelligence (AI), Crutzen and Hein (2009) propose the creation of Critical Transformative Rooms "where human and artificial actors can negotiate" a range of things such as "about the change of meanings" and "the approach to deal with the users' emotions" pertaining to technology related elements such as "frozenness, fear, doubt, and trust." Shneiderman (2020) advances the Human-Centered Artificial Intelligence (HCAI) framework in support of methods that "are more likely to produce designs that are Reliable, Safe & Trustworthy (RST)." Lock, Bednarz, Leao, and Pettit (2020) call for a reframing of urban dashboards to become more participatory as "part of a wider ecosystem of urban data platforms" while identifying "the still unexploited prospects of visualization" going "beyond traditional web and desktop displays" for more interactive and real-time uses. Mühlhaüser, Meurisch, Stein, Daubert, Von Willich, Riemann, and Wang (2020) propose Street Lamps as a Platform as a "strategically augmented" and "key enabling technology for smart cities" with a "governance and stakeholder perspective" involving citizens, cities and partnerships, government, and economy and business. Burns and Andrucki (2020) use discourse analysis to render visible the unpaid, low-paid, and volunteer labor, in the form of "digital care worker" that is "invisibilized in current research and public discourse" about smart cities.

Design Thinking

Pitsis et al. (2020) introduce a special issue on design thinking (DT) as an approach to accommodating the 4th Industrial Revolution, said to be emerging from the confluence of artificial intelligence (AI), 5G connectivity, and associated technologies such as the Internet of Things (IoT) where DT offers "a human-centered context and counterpoint." Beckman (2020) explores the question of whether to frame or reframe and where design thinking might best fit in relation to other research design approaches.

Experimentation, Exploration, and Experience

Schaffers et al. (2011) explain the Living Lab concept, as "a methodology and a model for organizing innovation programmes and projects and conducting innovation experiments." Friberg (2014) focuses on the concept of atmosphere in combination with aesthetic education in proposing an approach to the exploration of performing everyday practices involving the intersection of awareness and the sensorial and bodily in urban spaces. Demers and Potvin (2016) describe an ambience as involving "complexity because of the changing nature of the environmental conditions that generates it" such as "light, wind, sun, and sound" thus "creating varying distribution patterns of natural fluxes" and "people in the environment." For Demers and Potvin (2016), ambience research "suggests that adaptive opportunities and reinterpretation of existing ambiences could ultimately translate into new spaces to experience environmental delight for responsive inhabitants." Halegoua and Lingel (2017) employ multiple methods for exploring invisible networks in the form of "dark fiber" in order "to study infrastructural entanglements" as in, "the ways that digital infrastructures become embedded within physical and social environments, using "urban media archaeology" drawing on the work of Mattern (2015), as well as "mapping and fieldwork for exploring materialist histories and infrastructure ownership models" referencing the work of Parks (2013) and "ethnographies of infrastructure" drawing on the work of Star (1999) that may "account for different socio-cultural conditions that shape meaning, affect, and experience of infrastructure" drawing on the work of Larkin (2013).

Simulation, Urban Digital Twins, and VR. Schmitt (2015) advanced the use of simulation as an important method in urban planning and design to "test assumptions" and "visualize the results of the design over time" in urban regions. Morello and Piga (2015) point to the "increasing interest in simulation by society and the profession" from an architecture perspective, highlighting the question of "how simulation technologies will enable us to pay more attention to the intangible values of the ambiance of spaces." Piga, Morello, and Salerno (2017) explore urban simulation from interdisciplinary and practical perspectives in relation to urban design and urban representation. While researchers promote the use of simulation (Schmitt, 2015), Dembski et al. (2020) highlight the use of "urban digital twins" as "data models allowing collaborative processes" in smart cities. Dembski (Gauss Centre for Supercomputing, 2020) discusses limitations of simulation in relation to emotions where "joy and fear, the aesthetic experience of green space and architecture, and other emotion-driven factors" are said to "play a major role in the success of urban design but are very difficult to represent." Work by Feng (TU Delft, 2020) with virtual reality (VR) is being used to explore "how it may help explain the behaviour of pedestrians" along with "what motivates their choices" and other questions such as "how do people scan their environment?" Claiming "a complex socio-technical system" to be "an appropriate metaphor for modelling smart cities", Cunha, Rossetti, and Campos (2020) propose a "social simulation meta-model" for healthcare as an example, in response to the global COVID-19 pandemic. Given that "the world is becoming more densely urbanized", Crooks, Heppenstall, Malleson, and Manley (2021) describe and advance agent-based modeling as a simulation technique for smart cities as complex systems. Crooks et al. (2021) use this technique "to model social systems from the bottom up" in "the creation of artificial worlds in which individuals are given unique behaviors and rules" in order to "interact with each other and their environment" and "through such interactions" it is said that "more micro-patterns emerge" from "how individuals form crowds" to people and their movements in relation to work that "result in traffic jams" and "by integrating geographic information into such models" can be made to "turn abstract artificial worlds to those that mimic real-world locations." The process of "dynamic data assimilation" is noted as a "way to prevent a simulation from diverging from reality" so as "to occasionally incorporate more up-to-date data and adjust the model accordingly" as in, using "streaming data produced through near real-time observational datasets" from "social media or vehicle routing counters." According to Crooks et al. (2021), combining "data assimilation methods" with "agent-based models" may prove to "be transformative for the ways that some systems" such as "smart cities, are modeled."

Space Syntax, Public Spaces, and Contexts

According to Van Nes (2014), "space syntax deals with place *structure* and not with place *character*" providing measures for "the two primary all-to-all (all street segments to all other) relations." As such, the space syntax method is said to provide "a useful tool for comparison of the spatial changes in a before-and-after situation of structural urban changes in an area" where "what matters is the degree of accessibility, visibilities, adjacencies, openness and enclosures in built environments." Konomi and Sasao (2018) explore urban sensing approaches through "techniques and tools for collecting detailed behavioral data in large public spaces" where researchers employ technologies enabling them to "act according to mobile notifications." This approach enables researchers to "explore the challenges of human in the loop sensing" along with the design of "a mobile behavior sampling tool based on smart notifications to address the challenges of in-situ sampling" for spatial analytics. Boonen and Lievens (2018) explore live prototypes as a proxy technology for accommodating a more effective methodology for the "complex context of use" in smart cities such as "serendipity" and "unforeseen types of use". Farina, Kotsopoulos, and Casalegno (2018) identify "hybrid connected spaces" as "new contexts of use and behavior" enabled by the "almost symbiotic bond" underlying the physical and digital. Also concerned with space in relation to urban peripheries and extended urbanization, Connolly, Keil and Harris Ali (2020) address concerns with the density of urban areas in relation to disease, proposing the use of interdisciplinary "landscape political ecology approaches."

Self-Relevant Research

While self-relevant research is defined as "immersive field research on a topic with which the researcher has had significant personal experience, and which is important to or part of the researcher's self-identify" (Amabile and Hall,

2019), it could be argued that such a research approach could be very useful to the study of smart cities, taking people and their needs into consideration since, as Amabile and Hall (2019) claim, "*personal experience may actually be essential for truly valid research.*" Amabile and Hall (2019) note that, "we are having life experiences that parallel those of many of our participants in multiple ways" and that such "self-relevance, we believe, enables us to better understand their experiences" and possibly, "could lead to more valid discoveries, richer theories, and more interesting, persuasive publications."

3.2 Summary

Table 1 provides a summary by year and author for methods and their characteristics for use in the context of visibilities and invisibilities in smart cities and regions covering the timespan of 2009 to 2021. Crutzen and Hein (2009) propose negotiation spaces for human and artificial actors as critical transformation rooms. Gehl (2010) highlighted the importance of walking in the city as much more than walking. Appreciative inquiry has been used in city contexts for addressing problems as opportunities (ICMA Press, 2011) and more recently as a life-based/strength-based approach to change management (Grieten et al., 2018). Living Labs emerge as a model for innovation experiments (Schaffers et al., 2011), while Friberg (2014) uses performing everyday practices as an approach to atmospheres and aesthetic education, and Halegoua and Lingel (2017) employ multiple methods including ethnography, fieldwork, and mapping to uncover the infrastructural entanglements of "dark fiber" networks. Patel (2017) describes wonder as frame, spectacle, and interruption in architectural spaces. Farina et al. (2018) describe the integral bond of the physical and digital as "almost symbiotic" enabling "new contexts of use and behavior." Burns and Andrucki (2020) render visible through discourse analysis those rendered invisible in unpaid or low-paid digital care work. Van Es (2014) advanced the notion of space syntax enabling before and after comparisons of place structure in urban environments. Simulation is used by a number of researchers for working with the intangible value of the ambience of spaces (Morello & Piga, 2015); by Schmitt (2015) to test assumptions and visualize the results of designs over time; by Piga et al. (2017), pertaining to interdisciplinary urban design and representations, while Dembski et al. (2020) identify limitations of simulation in relation to emotions. Ambiences are explored by Demers and Potvin (2016) as experiential and adaptive opportunities while live prototypes are explored in relation to serendipity and contexts of use by Boonen and Lievens (2018). Digital art is used by Chevalier (2018) to explore and imagine cities virtually. Urban mapping is used by Dovey et al. (2018) to explore the city as a space for the imagination and for possibility. Farina et al. (2018) use hybrid connected spaces to explore contexts of use and behavior while (Konomi and Sasao, 2018) explore urban sensing technologies in large public spaces involving "humans in the loop." Self-relevant research as articulated by Amabile & Hall (2019) may be extensible to the study of smart cities since many urban researchers are living in and experiencing their city as part of their personal experience of everyday life. Intuitive inquiry is a qualitative method advanced by Anderson (2019) because of its' "spaciousness and permission" and it encourages "new ways of being human in a troubled world." Design thinking (DT) is highlighted by Beckman (2020) for framing and reframing while Pitsis et al. (2020) describe the human-centered focus of DT. Connolly et al. (2020) propose landscape political ecology as an approach in addressing concerns with density in relation to disease. Cunha et al. (2020) propose social simulation for agent-based modeling of healthcare to "provide useful insights allowing for the identification of key social practices influencing its performance." Schniderman (2020) advances the HCAI framework as a method for designs that are reliable, safe and trustworthy. Feng (TU Delft, 2020) explores the use of virtual reality (VR) in simulation to learn more about pedestrian behaviour. Van Es and De Lange (2020) advance datawalking as a research method for addressing challenges associated with urban data and the computational turn in order to enable people to become more attentive and attuned, finding a space for reflection, and for meaningfully situating data processes in everyday life. Mühlhaüser et al. (2020) propose Street Lamps as a Platform as novel infrastructure and enabling technology for smart cities. Lock et al. (2020) propose the evolving of urban dashboards into more participatory, platform spaces supporting interactive and real-time use. Crooks et al. (2021) propose the use of agent-based modeling as a simulation technique combined with data assimilation.

Across this array of characteristics associated with methodologies for InVisibilities in smart cities, context and imagination appear for more than one method as does the importance of people. Also of note is the presence of emotion, attuned, and ambiance. The terms in Table 1 inform and guide the selection of elements or proxies that are used in exploring relationships for InVisibilities in smart cities in this chapter.

Table 1. Overview of elements characterizing methodologies for InVisibilities in smart cities

Methodologies and their Characteristics for Smart Cities			
Author(s)	**Year**	**Methods**	**Characteristics**
Crutzen & Hein	2009	Critical Transformative Rooms	Negotiation by human and artificial actors
Gehl	2010	Walking	Information, experience, platform, framework
ICMA Press	2011	Appreciative Inquiry	City context for problems as opportunities
Schaffers et al.	2011	Living Labs	Model for innovation experiments
Friberg	2014	Performing everyday practices	Atmospheres & Aesthetic Education
Van Es	2014	Space Syntax	Before and after comparisons
Morello & Piga	2015	Simulation	Intangible value of the ambiance of spaces
Schmitt	2015	Simulation	Test assumptions; Visualize results over time
Demers & Potvin	2016	Ambiences	Experiential and adaptive opportunities
Halegoua & Lingel	2017	Ethnography, Fieldwork, Mapping	Experience of infrastructure
Patel	2017	Wonder	Frame, Spectacle, Interruption
Piga et al.	2017	Simulation	Interdisciplinary urban design & representation
Boonen & Lievens	2018	Live Prototypes	Serendipity, contexts of use
Chevalier	2018	Digital Art	Imagination
Dovey et al.	2018	Urban Mapping	Cities as spaces of possibility; Imagination
Farina et al.	2018	Hybrid connected spaces	Contexts of use and behavior
Grieten et al.	2018	Appreciative Inquiry	Life-based/strength-based
Konomi & Sasso	2018	Sensing	Humans in the Loop
Amabile & Hall	2019	Self-relevant research	Personal experience
Anderson	2019	Intuitive inquiry	Auspicious bewilderment
Beckman	2020	Design Thinking (DT)	Frame or reframe
Burns & Andrucki	2020	Discourse Analysis	Digital workers and un/low-paid caring
Connolly et al.	2020	Landscape political ecology	Density in relation to disease
Cunha et al.	2020	Social Simulation	Agent-based modeling
Dembski et al.	2020	Simulation	Design limitations in relation to emotions
Lock et al.	2020	Urban Dashboards	Reframing for interactive, real-time use
Mühlhaüser et al.	2020	Street Lamps as a Platform (SLaaP)	Strategically augmented
Pitsis et al.	2020	Design Thinking (DT)	Human-centered
Schniderman	2020	HCAI	Reliable, Safe, Trustworthy
TU Delft	2020	Simulation	Virtual Reality (VR)
Van Es & De Lange	2020	Datawalking	Attentive/Attuned, Reflective, Situated
Crooks et al.	2021	Simulation	Agent-based modeling + Data Assimilation

Figure 1. Overview of methodologies for visibilities and invisibilities in smart cities

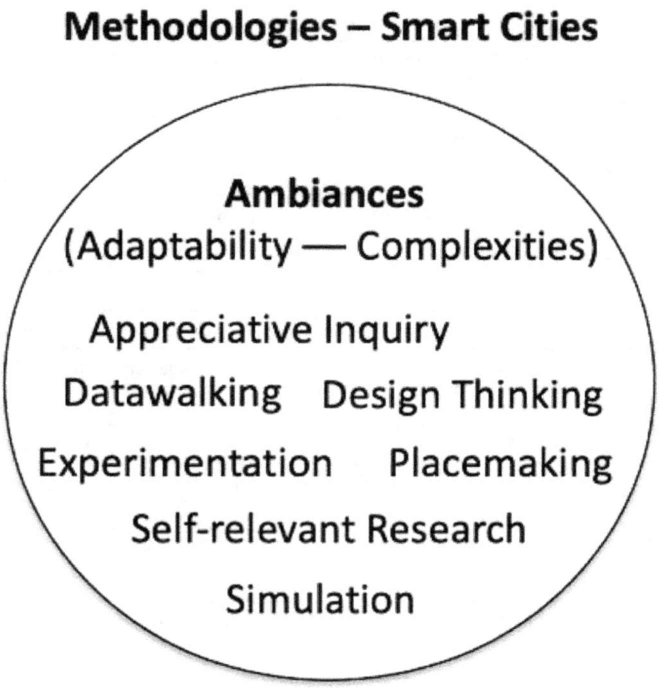

Graphically, this review of the research literature is depicted in Fig. 1 showing methodologies for working with visibilities and invisibilities in smart cities and regions.

Methodologies for smart cities are centered in this chapter around ambiances, characterized by adaptability and complexity, and include, but are not limited to, appreciative inquiry, datawalking, design thinking, experimentation (including exploration and experience), self-relevant research, and simulation.

3.3 InVisibilities and the Ambient: New Spaces for Methodologies in Smart Cities

Based on the review of the research literature presented in this chapter, a conceptual framework for methodologies accommodating spaces for visibilities and invisibilities in smart cities is advanced as depicted in Figure 2.

The interweaving of visibilities and invisibilities in smart cities focusing on characteristics of the ambient opens the way for novel approaches to inquiry, as

Figure 2. Methodologies accommodating spaces for visibilities and invisibilities in smart cities

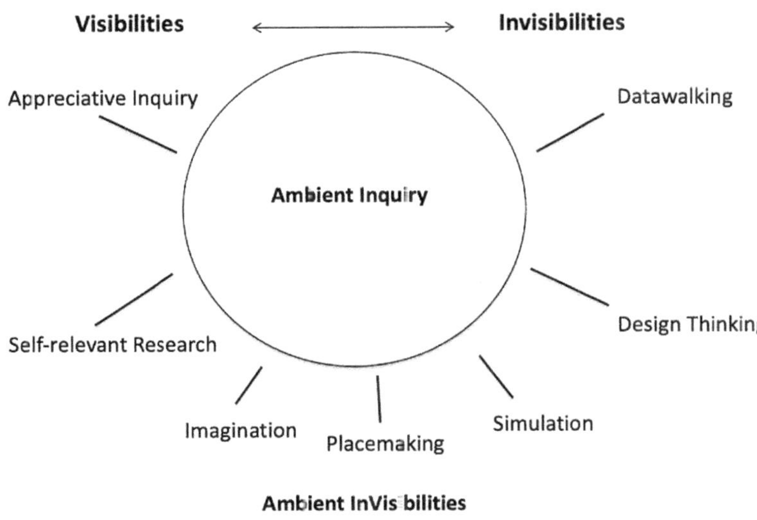

in, ambient inquiry accommodating appreciative inquiry, datawalking, design thinking, placemaking, imagination, self-relevant research, and simulation, to name a few, in support of the notion of ambient InVisibilities.

This framework contributes further to guidance for exploration of the research question for this chapter presented in Section 1 and reformulated here as a proposition, as follows.

P1: Ambient methodologies matter for cities, urban regions, and smart cities in accommodating explorations and implementations of emergent and evolving spaces for an interweaving of visible and invisible dimensions as InVisibilities in relation to the ambient.

How methodologies accommodate visibilities and invisibilities is explored in this chapter in the context of smart cities, and more particularly the ambient, featuring evermore awareness and attuning. Elements characterizing the ambient are explored (e.g., more complex, dynamic, in-the-moment, and adaptive) involving technologies, interactions, and people and their experiences and assessments of smart cities.

4. METHODOLOGY FOR EXPLORING AMBIENT APPROACHES

Based on the perspectives emerging from the review of the research literature in this chapter, methodologies emerge for exploring approaches to visibilities and invisibilities as InVisibilities in smart cities in relation to the ambient. The proxies of *placemaking* and *attuning to urban spaces* are used in this exploration of ambient inquiries for *InVisibilities* and *the ambient*, respectively.

4.1 Research Design for Exploring Ambient Types of Inquiries in Smart Cities

This chapter uses an exploratory case study approach, said to be particularly appropriate for the study of contemporary phenomena (Yin, 2018), which in this work focuses on *placemaking* as a proxy for approaches and methods pertaining to visibilities and invisibilities as *InVisibilities* and *attuning to urban spaces* as a proxy for the *ambient* in smart cities and regions. An explanatory correlational design is also incorporated into the research design in this chapter, as this provides "complementarity of case study and statistical research" (Yin, 2018). The study upon which this chapter is based was described using a website where, during sign up, demographic data were collected including elements such as age range, location (city), and people could self identify in one or more categories (e.g., community member, city official, researcher, educator, student, business, etc.). Upon sign up, people were invited to complete a survey about their experience of their city as a smart city. An invitation was also provided to participate in an in-depth interview about smart cities, online or in person, depending upon location. The study attracted participation from across Canada and extended to other cities in the United States, Ireland, and Israel, to name a few. The survey instrument, containing a mix of closed and open-ended questions, and the interview protocol were pretested prior to use in the study. To enable responses from a broad spectrum of people, heterogeneity sampling, a type of purposive sampling was used whereby an invitation to participate in the study was posted to webspaces with the potential to attract a wide response. In parallel with this study, data were also gathered systematically from a wide range of people across many sectors in small to medium to large-sized cities in Canada (e.g., Toronto, Vancouver, and greater Victoria) through individual and group discussions on a community level. Data analysis included the use of content analysis for qualitative data and descriptive statistics for quantitative data. The Real Statistics Software Resource Pack Add In (Zaiontz, 2019) for Microsoft Excel was used, consisting as it does of a wide range of tools for statistical analysis, such as the correlation feature for calculation of the Spearman correlation for the ordinal data, generated by assessments from survey respondents. The survey and interview are designed to enable people to assess their experience of the city and region where they live in a wide variety of ways including placemaking and attuning to urban spaces. Overall, data were analyzed for n=78 including 41% females and 59% males for people ranging in age from their 20s to their 70s.

4.2 Exploring InVisibilities through an Ambient Methodology Approach

When asked to assess to what extent do you associate smart cities with elements such as placemaking on a scale of 1 (Not at all) to 7 (absolutely), as indicated in Table 2, 17% responded at position 5 on the 7-point Likert-type scale for *placemaking*, 33% at position 6, and 50% at the upper end of the scale at position 7. When asked to assess whether city-focused social media and other aware technologies give rise to many possibilities, for *attuning to urban spaces*, 17% responded at position 4 on the scale, 33% at position 6, and 50% at position 7. The correlation between *InVisibilities* using *placemaking* as a proxy and *the ambient* using the proxy of *attuning to urban spaces* as ordinal data, was conducted using a feature of the Real Statistics Software Add In (Zaiontz, 2019) for Microsoft Excel showing the Spearman correlation coefficient result of .43 in Table 2.

It should be noted that placemaking is used as a proxy for visible and invisible dimensions of smart cities, interwoven as InVisibilities because it is concerned with the physical and visible dimension of place while also being concerned with the less visible elements of place that may pertain to emotion/affect, collaboration and the social, to name a few (PPS, 2007).

Looking at the scatterplot for this relationship, as placemaking (InVisibilities) increases or decreases, so too does attuning to urban spaces (the ambient), as depicted in Fig. 3.

Findings that emerged of a qualitative nature pertaining to methods and approaches involving *placemaking* and *attuning to urban spaces* based on interviews, open-ended survey responses, and group and individual discussion

Table 2. Assessments for InVisibilities (placemaking) and the Ambient (attuning) in smart cities

Items	Assessments	Correlation (Spearman)
InVisibilities (placemaking)	17% (5); 33% (6); 50% (7)	.43
The Ambient (attuning to urban spaces)	17% (4); 33% (6); 50% (7)	

Figure 3. Scatterplot for placemaking (InVisibilities) and attuning (the ambient) in smart cities

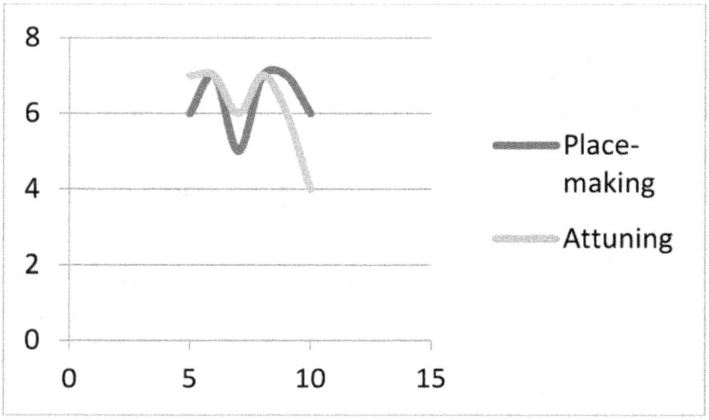

are presented below in relation to placemaking, design thinking, and data visualizations.

Placemaking

An educator "interested in the concept of place" suggested the term "wise place" as in, "make it friendly, comfortable" so that the smart city "reflects human values and needs" and this could involve "place based projects" at "any academic level" where the "student can make [a] real contribution." The educator suggested that such projects "mean that schools open their 'walls'." Another educator identified an initiative involving the use of a mobile app "that people could use to capture ideas, evidence of problems on the move" allowing also for "evidence of potential solutions and impact" in keeping it would seem with notions of datawalking (Van Es and De Lange, 2020). A community member stressed the need for "educating people" about the smart city concept as in, "why it's important, what opportunities exist, and how to make it about more than data" in order "for the general public to be interested and understand more." As if in response, a community leader identified the use of digital platforms (e.g., PlaceSpeak) as a way to "encourage people to do some observation of space and people" in order to "look at any given space and analyze what's working and what isn't and what could improve it" and this would seem to pertain to the importance of contexts as discussed by Van Es and De Lange (2020), Farina et al. (2018), and Boonen and Lievens (2018). A community leader identified the need for "finding fun and meaningful ways to engage the public" as a key element contributing to the idea of a smart city.

Design Thinking

Design thinking was used during an urban initiative to involve people across the city in providing guidance and ideas for reusing and recycling, as in upcycled urbanism, for portions of an iconic, historic infrastructural urban element about to be decommissioned. The initiative attracted cross-boundary collaborations with placemakers, city officials, architects, community leaders and members, to name a few. It is worth noting that live prototyping, reminiscent of work by Boonen and Lievens (2018), was used during the initiative to articulate and share ideas. Imagination, storytelling scenarios, and visualizations were also used in describing a range of ideas for consideration by the city. Other ideas proposed included the constructing of a physical platform involving counterweights, stairs, and climbing elements for viewing and experiencing. Ideas were also developed and shared involving the use of lights, lasers, screen projections, and other artistic elements for visualizations and visual renderings involving culture, history, and learning, featuring urban transformations.

Data Visualizations

Regarding urban data in smart cities, a community leader identified the need to "visualize this data, and 'simplify' it for the larger public." A city information technology (IT) manager stated that "we're starting to look at the tools to help us mine the data that we already have an interest in" indicating that "we're very much immature in that overall data sense" so "we're just starting to look at the tools that would give us the visualization." As an alternative to dashboards of data, an educator suggested that "you can take these bits of data and rather than just present numbers you can make beautiful artistic visualizations" for projection "on the walls" as art in public spaces "to make beautiful pictures, patterns" and the like. From the perspective of public administration, the need for a "scoreboard" was suggested to identify "milestones as strategy is deployed" in steps toward smart city development and implementation. On a digital level, a public administrator proposed, "participating in platforms of sharing information" using the example of travel and traffic. Also, a public administrator identified key elements for smart cities in terms of the need for "learning via its interconnections with other cities", "learning capabilities", and "a learning strategy."

4.3 Discussion and Interpretation of Findings

While the correlation found in this work between visible and invisible dimensions of smart cities as *InVisibilities* using *placemaking* as a proxy and *the ambient* using *attuning to urban spaces* as a proxy is .43, Creswell (2018) indicates that correlations above .35 "are useful for limited prediction" and "are the typical values used to identify variable membership in the statistical procedure of factor analysis" as in "the intercorrelation of variables." As such, early-stage findings in this chapter serve to advance the potential of ambient approaches in evolving understandings of smart cities while opening spaces for exploring other variables and their interconnections.

Through approaches such as placemaking and design thinking, it is worth noting the emergence of platforms as both physical and digital, highlighting perhaps the intermingling of the visible and the invisible, for the viewing and sharing of information, art, history, and culture, contributing to learning. Findings associated with moving beyond number-driven urban dashboards to more meaningful, artistic visualizations of urban data also support the framing and reframing of urban dashboards proposed by Loch et al. (2020). Approaches such as placemaking or data visualization would seem to enable spaces to accommodate "smart cities" as "wise cities" taking into consideration the needs of people for fun, meaningfulness, and learning, among many other things.

5. EMERGING RESEARCH AND OPPORTUNITIES

Emerging trends are identified in this chapter for practice and research in sections 5.1 and 5.2 pertaining to InVisibilities and the ambient through approaches and methodologies in smart cities and regions.

5.1 Practice

In terms of practice, this chapter contributes to spaces for the rethinking of methods and approaches in addressing challenges associated with the interweaving of the visible and invisible in smart cities and to emergent and associated understandings of the ambient.

Ambient InVisibilities. This chapter provides insight into understandings of ambient approaches as more adaptive, dynamic, and community-based approaches for smart cities pointing to emerging potentials associated with placemaking, design thinking, and data visualizations in working with the visible and invisible. Emphasis is placed on enabling broad, cross-sector participation in the development, design, and implementation of initiatives that are more responsive to contemporary urban challenges and opportunities for complex problem-solving. As such, areas for exploration potential include:

1. *Live Prototyping* fostered through more dynamic urban initiatives that may involve design thinking while involving educational institutions, communities, and local government
2. *Visualizations* fostered through the gathering, sharing and use of more dynamic urban data initiatives that may involve educational institutions, artists, designers, developers, community members, and local government

5.2 Research

This chapter advances the potential for future research opportunities and emerging discourse spaces for urban methodologies as ambient in the context of smart cities.

Ambient Methodologies and Spaces

The viability of the conceptual framework for methodologies accommodating spaces for visibilities and invisibilities in smart cities is operationalized for use in this chapter, demonstrating early-stage potentials while offering and opening up possibilities for the evolving of urban research initiatives for testing and practice.

Ambient InVisibilities

Articulation of spaces for real-time, dynamic, and adaptive types of approaches to realizing the interactive and interwoven nature of the invisible and the visible emerge in this chapter opening the way for evolving methodologies and possibly novel opportunities for collaborations involving citizens, businesses, researchers, students, and governments. As such, areas for exploration potential include:

1. *Education* fostered through more dynamic urban initiatives involving educational institutions, communities, and local government
2. *Learning* about smart cities fostered through more dynamic and meaningful real world data collection and analysis activities using mobile apps and other urban platforms
3. *Engagement* fostered through more dynamic and adaptive urban activities and data visualizations that are fun and meaningful

Ambient Approaches and Ambient Inquiry

This chapter opens the way to evolving research opportunities and challenges related to the potential for movement toward the building of more dynamic and in-the-moment inquiry opportunities such as ambient inquiry as an urban research technique with implications for practice.

6. CONCLUSION

This chapter opens spaces for consideration of placemaking and the ambient as dimensions of visibilities and invisibilities for smart cities in relation to what is emerging from interdisciplinary perspectives involving the innovating of spaces for urban methodologies. Placemaking, design thinking (DT), and data visualization emerge as key methodologies for navigating invisibilities and visibilities as InVisibilities and the ambient in smart environments. Spaces for these methodologies will be enabled through fostering increased education and adaptive thinking. The main contributions of this chapter include a) articulation and operationalization of a conceptual framework for methodologies accommodating spaces for visibilities and invisibilities in smart cities; b) further development of the research literature for visibilities and invisibilities in smart environments by the extending of dimensions to include *InVisibilities* and *the ambient* represented in this work by the proxies of *placemaking* and *attuning to urban spaces*; and c) highlighting the elements of education, learning, and engagement in fostering smarter, more responsive, cities and regions.

Limitations of this chapter pertain to the small sample size of the study upon which this exploration is based. An additional limitation of this chapter may be related to the use of proxies for InVisibilities (e.g., placemaking) and the ambient (e.g., attuning to urban spaces) and this is mitigated by actual examples from the research literature and in-depth discussions with a range of individuals regarding everyday emergent practices. These limitations could also be further mitigated going forward by exploring InVisibilities and the ambient more directly or using other variables. Other limitations related to literature review domain coverage are mitigated by the interdisciplinary potential to extend this review to other areas going forward.

This chapter continues to give form, detail, and definition to the exploration of dimensions of visibilities and invisibilities in smart cities and regions along with methodologies to inform research and practice. The analysis of findings identifies a positive correlation of .43 between *placemaking* and the *ambient*, constituting what Creswell (2018) describes as "limited prediction" potential, important for exploring the "intercorrelation of variables" going forward in smart cities. This finding would seem to give support to the proposition under exploration in this chapter, that ambient methodologies matter for cities, urban regions, and smart cities in the context of visible and invisible dimensionalities.

Insights of particular interest to the reader include the following:

1. The emerging range of dimensions for invisibilities and visibilities in the form of *placemaking* as *InVisibilities* and *attuning to urban spaces* as *the ambient* in smart cities
2. The presence of "limited prediction" emerged when *InVisibilities* (e.g., *placemaking)* and *the ambient* (e.g., *attuning to urban spaces)* were correlated in a smart city context
3. The interplay of datawalking, data visualizations, and design thinking as approaches to *InVisibilities* and *the ambient*

Ideas highlighted in this chapter pertain to the following:

1. The importance of people becoming more involved in understandings of *placemaking* and *the ambient* as dimensions of invisibilities and visibilities in smart cities
2. Potentials for people to learn, educate, and engage through *placemaking* and *attuning to urban spaces*
3. Explorations of *placemaking* and *the ambient* as novel research and practice spaces informing approaches and methodologies for smarter cities and regions

Exercises for the reader to consider, encouraging further exploration include the following:

1. Identify one or more examples of *InVisibilities* and *placemaking* in your city
2. Identify one or more examples of *the ambient* and *attuning to urban spaces* in your city
3. Describe why *InVisibilities* in the form of *placemaking* and *attuning to urban spaces* could be important for smart cities, based on your example(s)

If you would like to share your responses to the Exercises with the author, please complete the questions in this online space for Chapter 6 – InVisibilities and Methods

This work will be of interest to a broad audience including students, educators, researchers, artists, architects, data scientists, urban planners, urban practitioners, community members, public administrators, and anyone concerned with emerging spaces and methodologies that take into consideration the complex urban challenges and opportunities posed in exploring renderings of the invisible and visible in smart, responsive, and future cities.

REFERENCES

Amabile, T. M., & Hall, D. T. (2019). The undervalued power of self-relevant research: The case of researching retirement while retiring. *Academy of Management Journal*. Advance online publication. doi:10.5465/amp.2018.0083

Anderson, R. (2019). Intuitive inquiry: Inviting transformation and breakthrough insights in qualitative research. *Qualitative Psychology*, *6*(3), 312–319. doi:10.1037/qup0000144

Beckman, S. L. (2020). To frame or reframe: Where might design thinking research go next? *California Management Review*, *62*(2), 144–162. doi:10.1177/0008125620906620

Boonen, M., & Lievens, B. (2018). The use of live-prototypes as proxy technology in smart city living lab pilots. In DAPI 2018 (LNCS 10921, pp. 203-213). doi:10.1007/978-3-319-91125-0_17

Burns, R., & Andrucki, M. (2021). Smart cities: Who cares? *Environment and Planning A. Economy and Space*, *53*(1), 12–30. doi:10.1177/0308518X20941516

Chevalier, M. (2018). *Meta-Cities 7, Ubiquity 1 & 2*. Retrieved 22 April 2018 from https://www.mayorgallery.com/exhibitions/526/overview/

Chevalier, M. (2020). *Biography*. Retrieved 13 May 2020 from https://www.miguel-chevalier.com/en/biography

Connolly, C., Keil, R., & Harris Ali, S. (2020). Extended urbanisation and the spatialities of infectious disease: Demographic change, infrastructure and governance. *Urban Studies (Edinburgh, Scotland)*, 1–19. doi:10.1177/0042098020910873

Creswell, J. W. (2018). *Educational Research: Planning, Conducting, and Evaluating Quantitative and Qualitative Research* (6th ed.). Pearson.

Crooks, A. T., Heppenstall, A., Malleson, N., & Manley, E. (2021). Agent-Based Modeling and the City: A Gallery of Applications. In W. Shi, M. Goodchild, M. Batty, M.-P. Kwan, & A. Zhang (Eds.), *Urban Informatics* (pp. 885–910). Springer. doi:10.1007/978-981-15-8983-6_46

Crutzen, C. K. M., & Hein, H.-W. (2009). Invisibility and visibility: The shadows of artificial intelligence. In *Handbook of Research on Synthetic Emotions and Sociable Robotics: New Applications in Affective Computing and Artificial Intelligence*. IGI Global. doi:10.4018/978-1-60566-354-8.ch024

Cunha, M. E. S., Rossetti, R. J., & Campos, P. (2020). Modelling Smart Cities Through Socio-Technical Systems. In *2020 IEEE International Smart Cities Conference (ISC2)* (pp. 1-8). IEEE. 10.1109/ISC251055.2020.9239042

Delft, T. U. (2020). Evacuating virtual building. *TU Delft Stories*. Retrieved 3 June from https://www.tudelft.nl/en/stories/articles/evacuating-virtual-buildings/

Dembski, F., Wössner, U., Letzgus, M., Ruddat, M., & Yamu, C. (2020). Urban digital twins for smart cities and citizens: The case study of Herrenberg, Germany. *Sustainability*, *12*(6), 2307. doi:10.3390u12062307

Demers, C., & Potvin, A. (2016). From history to architectural imagination: A physical ambiences laboratory to interpret past sensory experiences and speculate on future spaces. *Ambiances: International Journal of Sensory Environment, Architecture and Space*, 756.

Dovey, K., Pafka, E., & Ristic, M. (Eds.). (2018). *Mapping urbanities: Morphologies, flows, possibilities*. Routledge.

Farina, C., Kotsopoulos, S. D., & Casalegno, F. (2018). Hybrid connected spaces: Mediating user activities in physical and digital spaces. In NDAPI 2018 (LNCS 10921, pp. 35-55). Academic Press.

Friberg, C. (2014). Performing everyday practices: Atmosphere and aesthetic education. *Ambiances: International Journal of Sensory Environment, Architecture and Space*, 464.

Gauss Centre for Supercomputing. (2020). *Using digital twins to design more sustainable cities: HLRS uses supercomputing and visualization to develop comprehensive models of urban environments, supports city planning in Herrenberg*. EurikaAlert. Retrieved 13 May 2020 from https://www.eurekalert.org/pub_releases/2020-05/gcfs-udt050720.php

Gavalas, D., Nicopolitidis, P., Kameas, A., Goumopoulos, C., Bellavista, P., Lambrinos, L., & Buo, B. (2017). Smart cities: Recent trends, methodologies, and applications (Editorial). Wireless Communications and Mobile Computing.

Gehl, J. (2010). *Cities for people*. Island Press.

Grieten, S., Lambrechts, F., Bouwen, R., Huybrechts, J., Fry, R., & Cooperrider, D. (2018). Inquiring into appreciative inquiry: A conversation with David Cooperrider and Ronald Fry. *Journal of Management Inquiry*, *27*(1), 101–114. doi:10.1177/1056492616688087

Halegoua, G., & Lingel, J. (2017). Invisibility and hypervisibility: Imagining publics for urban broadband networks. *The 18th Annual Conference of the Association of Internet Researchers*.

ICMA Press. (2011). Positive problem solving: How appreciative inquiry works. *InFocus: Strategies and Solutions for Local Government Managers, 43*(3). Retrieved 11 May 2020 from http://www.southcentral.edu/webdocs/research_and_planning/Strategic_Planning/7%20Appreciative%20Inquiry%20Articles/Positive%20Problem%20Solving%20sallie%20Lee.pdf

Konomi, S., & Sasao, T. (2018). Designing a mobile behavior sampling tool for spatial analytics. In DAPI 2018 (LNCS 10922, pp. 92-100). doi:10.1007/978-3-319-91131-1_7

Larkin, B. (2013). The politics and poetics of infrastructure. *Annual Review of Anthropology, 42*(1), 327–343. doi:10.1146/annurev-anthro-092412-155522

Lock, O., Bednarz, T., Leao, S. Z., & Pettit, C. (2020). A review and reframing of participatory urban dashboards. *City, Culture and Society, 20*, 100294. Advance online publication. doi:10.1016/j.ccs.2019.100294

Mattern, S. (2015). *Deep mapping the city*. University of Minnesota Press. doi:10.5749/9781452958491

Mehl, M. R., & Connor, T. S. (2012). *Handbook of Research Methods for Studying Daily Life*. The Guilford Press.

Morello, E., & Piga, B. E. A. (2015). Experiential simulation in architecture and urban space: Introduction to the special issue. *Ambiances: International Journal of Sensory Environment, Architecture and Space*, 671.

Mühlhaüser, M., Meurisch, C., Stein, M., Daubert, J., von Willich, J., Rieman, J., & Wang, L. (2020). Street lamps as a platform. *Communications of the ACM, 63*(6), 75–83. doi:10.1145/3376900

Parks, L. (2013). Earth observation and signal territories: Studying U.S. broadcast infrastructure through historical network maps, Google Earth, and fieldwork. *Canadian Journal of Communication, 38*(3), 285–307. doi:10.22230/cjc.2013v38n3a2736

Patel, A. (2017). Wonder as an interruption. *Conscious Cities*. Retrieved 23 April 2018 from https://www.ccities.org/wonder-as-an-interruption/

Piga, B. E. A., Morello, E., & Salerno, R. (2017). Introducing a research perspective in urban design and representation. In B. E. A. Piga & R. Salerno (Eds.), *Urban design and representation: A multidisciplinary and multisensory approach*. Springer. doi:10.1007/978-3-319-51804-6_1

Pitsis, T. S., Beckman, S. L., Steinert, M., Oviedo, L., & Maisch, B. (2020). Designing the future: Strategy, design, and the 4th industrial revolution – an introduction to the special issue. *California Management Review, 62*(2), 5–11. doi:10.1177/0008125620907163

PPS. (2007). *What is placemaking?* New York, NY: Project for Public Spaces. Retrieved 25 April 2021 from https://www.pps.org/article/what-is-placemaking

Schaffers, H., Komninos, N., Pallot, M., Trousse, B., Nilsson, M., & Oliveira, A. (2011). Smart Cities and the Future Internet: Towards Cooperation Frameworks for Open Innovation. The Future Internet, 6656.

Schmitt, G. (2015). *Information Cities.* https://itunes.apple.com/us/book/information-cities/id970529491?mt=11

Shneiderman, B. (2020). Human-Centered Artificial Intelligence: Reliable, Safe & Trustworthy. *International Journal of Human-Computer Interaction, 36,* 6. doi:10.1080/10447318.2020.1741118

Star, S. (1999). The ethnography of infrastructure. *The American Behavioral Scientist, 43*(3), 382–391. doi:10.1177/00027649921955326

Van Es, K., & De Lange, M. (2020). Data with its boots on the ground: Datawalking as research method. *European Journal of Communication*, 1–12.

Van Nes, A. (2014). Space syntax in theory and practice. In Geodesign by Integrating Design and Geospatial Sciences. Springer. doi:10.1007/978-3-319-08299-8_15

Yin, R. K. (2018). *Case study research and applications: Design and methods* (6th ed.). Sage.

Zaiontz, C. (2019). *Real Statistics using Excel.* www.real-statistics.com

ADDITIONAL READING

Bozikovic, A. (2020). After Sidewalk Labs, Toronto should get smart. *Globe and Mail.* Retrieved 8 May 2020 from https://www.theglobeandmail.com/canada/toronto/article-after-sidewalk-labs-toronto-should- get-smart/

CORDIS. (2020). Cities are becoming digital, thanks to the urban data platforms that enable it. *TechXplore*. Retrieved 20 May 2020 from https://techxplore.com/news/2020-05-cities-digital-urban-platforms-enable.html

De Matteis, F., Bille, M., Griffero, T., & Jelić, A. (2019). Phenomenographies: Describing the plurality of atmospheric worlds. *Ambiances,* 5. Retrieved 25 April 2020 from https://journals.openedition.org/ambiances/2526

Gedenk, E. (2020). Using digital twins to design more sustainable cities. *High-Performance Computing Center Stuttgart (HLRS) News*. Retrieved 8 May 2020 from https://www.hlrs.de/news/detail-view/2020-05-07/

Kaiying, C. L., Plewe, D. A., & Röcker, C. (2015). The ambience of ambient intelligence: An Asian approach to ambient systems? *Procedia Manufacturing*, *3*, 2155–2161. doi:10.1016/j.promfg.2015.07.355

Parisi, S., Holzbach, M., & Rognoli, V. (2020). The Hybrid Dimension of Material Design: Two Case Studies of a Do-It-Yourself Approach for the Development of Interactive, Connected, and Smart Materials. In T. Ahram, W. Karwowski, A. Vergnano, F. Leali, & R. Taiar (Eds.), *Intelligent Human Systems Integration 2020. IHSI 2020. Advances in Intelligent Systems and Computing* (Vol. 1131). Springer. doi:10.1007/978-3-030-39512-4_140

Wakefield, J. (2020). Coronavirus: Google ends plans for smart city in Toronto. *BBC News*. Retrieved 7 May 2020 from https://www.bbc.com/news/technology-52572362

Wiggins, J. (2016). How do smart cities talk to people? *Wilson Center, Urban Sustainability Laboratory*. Retrieved 11 May 2020 from https://www.wilsoncenter.org/article/how-do-smart-cities-talk-to-people

KEY TERMS AND DEFINITIONS

Ambient Methodologies: Ambient methodologies refer to more adaptive approaches to exploration of awareness in relation to people and awareness-enabled and enabling technologies encompassing the physical, digital, and less tangible, while being attentive to ambiences, atmospheres, augmentations, emotion/affect, and the like.

Ambient Spaces: Ambient spaces refer to more adaptive conceptualizations in support of awareness in relation to people and awareness-enabled and enabling technologies informing urban design involving the physical, digital, augmented, and the like.

Ambient Visibilities/Invisibilities: Ambient visibilities/invisibilities refer to urban elements accommodating more aware people and awareness-enabled technologies encompassing the physical, the digital, and the less tangible.

Attuning: Attuning refers to the concept or quality of being attuned and aware as it applies to people on the one hand, to technologies on the other, within physical and digital spaces and networks.

Invisibilities: Invisibilities refer to urban elements that are rendered invisible or that may present as intangibles.

Smart Cities: Smart cities are urban areas and regions characterized by more aware and engaged people, interacting with each other and aided by the use of more awareness enhancing technologies for a wide range of purposes from communication to design to atmospheres to mobility to livability, to governance, to data generation and use for learning and informed decision-making.

Visibilities: Visibilities refer to physical and other visible urban elements and infrastructures.

Section 4

Ambient Invisibilities Going Forward

Chapter 7
Ambient Approaches and Inquiries:
Emerging Opportunities for Smart Cities and Future Cities

ABSTRACT

The purpose of this chapter is to explore urban visibilities and invisibilities through the use of ambient approaches and inquires in smart cities and regions. As such, this chapter seeks to further operationalize elements of ambient InVisibilities advanced in this book, incorporating livability and well-being and challenges and opportunities associated with data. The research literature for well-being and livability in relation to urban data visibilities and invisibilities is explored in this chapter in the context of smart and responsive cities. Using an exploratory case study approach, combined with an explanatory correlational design, selected elements of urban life in smart cities are is investigated. This chapter makes a contribution to 1) the research literature for well-being and livability and other emergent complexities associated with data in the context of smart cities and 2) the evolving of theory through formulation of a conceptual framework for augmented awareness for ambient Invisibilities in smart cities.

DOI: 10.4018/978-1-7998-3850-0.ch007

1. INTRODUCTION

Acuto (2020) offers "lessons for urban(izing)" claiming that while "COVID-19 has magnified the deficiencies of how we manage our cities" and may "drive a change in the way we think about cities and health" the global pandemic is also providing "a unique chance to rethink, replan, and redesign." As such, concern regarding the implications for smart cities and regions of the global COVID-19 pandemic occurring during the writing of this work gives rise to opportunities for considering ambient approaches and inquires as a pathway to conducting some rethinking, replanning, and redesigning of smart cities. Smart cities in this work refer to urban areas and regions, characterized by more aware and engaged people, interacting with each other and aided by the use of more awareness enhancing technologies for a wide range of purposes including but not limited to well-being, livability, and data access and sharing.

This chapter employs an exploratory case study along with an explanatory correlational design where data are collected through the use of multiple methods including survey and in-depth interviews. In parallel with this data are also gathered systematically through group and individual discussions. The significance of this chapter pertains to the use of variables for well-being advanced for understanding smart cities in relation to visibilities and invisibilities, providing important insights as well as the potential for developing ambient approaches and inquiries. This chapter makes a contribution by a) extending the research literature for invisibilities and visibilities to include well-being and livability and other emergent complexities associated with data in the context of smart cities; and b) formulating a conceptual framework for augmented awareness for ambient InVisibilities in smart cities. As such, the objective of this chapter is described below.

Objective: The objective of this chapter is to explore and advance understandings of ambient approaches to inquiry in smart cities. The key research question posed is – *What is the nature of the relationship between the ambient and InVisibilities as well-being in smart cities and why does the ambient and well-being matter?*

2. BACKGROUND AND OVERVIEW

Keil (2020) points to "the threat of infectious disease in our planetary urban system" as possibly "an indication for how the world is connected" in a world of "extended urbanization" placing "limits to global urbanization" not to mention "challenges for urban planning" and yet, we are connected as well, digitally. Citing work by Connolly, Keil, and Ali (2020), Keil (2020) also notes that such extended urbanization is further challenged during this time of the COVID-19 global pandemic by "newfound closures, boundaries and limits that are imposed to make life liveable under global exposure." Where McCann (2020), drawing on the work of Berman (1982), uses invisibility and visibility to discuss COVID-19 in the city, as one of four dialectal tensions, it is worth noting the other three dialectal tensions as they may have a particular relevance to smart cities – privilege and privation; selfishness and solidarity; and absence and presence. Indeed, well-being and livability may assist in making the case further for use of methods such as self-relevant research (Amabile and Hall, 2019) while acknowledging attendant cautions identified by the researchers.

In an interview about the global pandemic (Watts, 2020), Latour observes that "the lockdown forced everyone into a kind of retreat, a moment of reflection" that "was quite extraordinary" giving rise to many questions and "a way of thinking about how they would create a better future." Hsiang, Allen, Annan-Phan, et al. (2020) note that "actions, such as closing schools and restricting populations to their homes, impose large and visible costs on society, but their benefits cannot be directly observed" so a study was conducted across six countries and "the researchers found that travel restrictions, business and school closures, shelter-in-place orders and other non-pharmaceutical interventions averted roughly 530 million COVID-19 infections across the six countries in the study period ending April 6" and that "Of these infections, 62 million would likely have been 'confirmed cases,' given limited testing in each country." As such, the researchers claim that "the deployment of anti-contagion policies" in the six countries studied "significantly and substantially slowed the pandemic." Articulated as a possibly transformative event "through our individual sacrifices, people everywhere" are said to "have each contributed to one of humanity's greatest collective achievements" adding that "I don't think any human endeavor has ever saved so many lives in such a short period of time" (Hsiang, 2020). The researchers noted the "huge personal costs to staying home and canceling events" while claiming that "the data show that each day made a profound difference" and "using science and cooperating, we changed the course of history" (Hsiang, 2020).

2.1 Definitions

Definitions are presented here, for key terms used in this chapter. Additional terms with definitions, as articulated by this work, are included in the *Key Terms and Definitions* section at the end of this chapter.

Ambient Theorizing: Chapter 5 of this work describes ambient theorizing as a framework for, and understanding of, smart cities taking aware people and aware technologies into consideration.

Ambient Inquiries: Chapter 6 of this work describes ambient inquiry in terms of building more dynamic and in-the-moment inquiry opportunities as an urban research technique with implications for practice.

2.2 Overview

This chapter provides a range of perspectives in Section 3 on ambient approaches and inquires in the context of visibilities and invisibilities in smart cities based on the research and practice literature, enabling formulation of a conceptual framework for operationalization and use in the exploration conducted in this chapter. Section 4 provides a description of the methodology used in this work, followed by the results, and a discussion of findings. Section 5 identifies opportunities associated with future directions for research and practice and Section 6 concludes with chapter coverage highlights, the major contributions, limitations, and key insights and ideas. Several exercise questions are also included for the reader in support of further exploration.

The primary audiences for this chapter include students, educators, researchers, policy makers, businesses, community members, and urban practitioners concerned with methods and theory for working with visibilities and invisibilities associated with well-being, livability, and data sharing in relation to urban design, urban planning, and other challenges and opportunities in the context of smart cities, responsive cities, and future cities.

3. PERSPECTIVES: AMBIENT APPROACHES AND INQUIRIES IN SMART CITIES

In the context of visibilities and invisibilities in smart cities, a review of the research literature is provided in relation to well-being and emergent complexities associated with urban data and livability, enabling a framing of ambient approaches and inquiries.

3.1 InVisibilities and Well-Being in Smart Cities

Anderson (2011) developed a body insight scale (BIS) to measure body awareness in "three forms" as "physical, psychological, and spiritual well-being" explored in this book in Chapter 2 in relation to feelings of tension in urban environments. Taking issue with the hype surrounding artificial intelligence (AI), Jordan (2019) argues that what is required is "well-thought-out interactions of humans and computers to solve our most pressing problems" with "computers to trigger new levels of human creativity, not replace human creativity." Jordan (2019) calls for "a wide range of voices" together with "perspectives" from the "social sciences, the cognitive sciences, and the humanities" and "the need for humans to understand and shape technology as it becomes ever more present and influential in their daily lives." Indeed, Jordan (2019) identifies the need for "the creation of a new branch of engineering" as in "an engineering discipline yet to emerge for the data- and learning-focused fields." Roitblat (2020) probes understandings of intelligence, artificial intelligence, and general artificial intelligence, claiming that "algorithms are not enough." Weekes (2020) describes a joint UK-Japan project focused on emotional AI in smart cities in relation to happiness and wellbeing and concerns with ethics, risks, and social implications "that arise from the use of different techniques that allow our bodies, feelings, intentions and emotional states to be read by machines." Amabile (2020) argues for the need to explore creativity and artificial intelligence (AI), defined as "the production of highly novel, yet appropriate, ideas, problem solutions, or other outputs by autonomous machines" giving rise to the question, "Will it [AI creativity] eventually produce novel sociotechnical systems that undermine human well-being?" Concerned with well-being and the quality of life for cyclists in smart cities, Oliveira, Costa, and Dias (2020) advance the notion of "environmental monitoring" focusing on factors such as "temperature, humidity, noise pollution, ultraviolet radiation, and light intensity" using "low-cost, mobile sensors-based devices attached to bikes" that provide "identification of the most recommended areas for cycling." Ullah, Al-Turjman, Mostarda, and Gagliardi (2020) provide a review of the role of artificial intelligence (AI), machine learning (ML), and deep reinforcement learning (DRL) in terms of "the evolution of smart cities" where details are provided for a range of applications from cybersecurity to energy-efficient utilization of smart grids (SGs) to smart health care systems. Lytras, Visvizi, Torres-Ruiz, Damiani, and Jin (2020) provide a space for the exploration of well-being in relation to urban computing in smart cities taking into consideration services, applications, and policymaking. From a machine learning (ML) perspective, Jordan identifies overuse of the term artificial intelligence (AI) (Pretz, 2021) and the need for engineering to incorporate greater consideration for people where "human happiness should not be an afterthought when developing technology."

3.2 Other Emergent Complexities and Smart Cities

Cities worldwide are facing emergent and evolving complexities associated with the global COVID-19 pandemic on a scale difficult to fathom. And yet, such complexities provide opportunities as witnessed by the work of researchers associated with BlueDot and the "outbreak science" software designed to detect and track virus outbreaks which worked in the case of the coronavirus outbreak in late 2019 (Whitaker, 2020). Weeden (2020) addresses the urban and rural divide in the time of the coronavirus pandemic in terms of the "right to go to a cottage", suggesting the need for "patience and the reimagining of the right to be rural as a civic responsibility." It is worth noting that Gao and O'Neil (2020) define urban land as "the Earth's surface that is covered primarily by manmade materials, such as cement, asphalt, steel, glass, etc." Paletta (2020) comments on airports as "expensive to build, hard to adapt to other uses, and now facing pandemic-related challenges" such that "airport terminals often live short, difficult lives." Paletta (2020) points to the "exponential increases in use" of airports followed more recently by "abruptly vanishing use" due to the coronavirus global pandemic "which will counterintuitively require more facility space to integrate social distancing and add yet-unbuilt health screening facilities."

In an interview with Michele Acuto of Connected Cities Lab (Klaus, 2020a) on "density, urbanization, and pandemic preparation", Klaus claims that "disease shapes cities" in a quick review of "urban planning and management" associated with past outbreaks. Acuto (Klaus, 2020a) calls for "a new imagination of the urban data we rely on" while acknowledging "the challenges of evidence that has different levels of legitimacy" giving way to "the need to rethink who are the right sources." Given the value of real time data for decision-making, Acuto (Klaus, 2020a) points to the importance of "digital infrastructure" as possibly "the sanitation of our time" in terms of pandemic responses in earlier times. Klaus (2020b) suggests that the post-pandemic future is already here even as countries are experiencing a second wave of the pandemic or, as is emerging, a third and even fourth wave. Schaller (2020) argues that, "Density isn't easy. But it's necessary" and that what is also needed is "the vision, commitment and fortitude to make our cities equitable, affordable and sustainable as well as dense, creative, and diverse." Where some are said to have sought out the edge of cities to escape urban density (Schaller, 2020), Acuto (2020), referring to the work of Connolly et al. (2020) "on the relative unmanageability of the suburban sprawl", reminds us "how coronavirus contagions can start and spread inward from, and between, the edges" of cities. From an architectural perspective, Chayka (2020) claims that, "the future of cities will be a fundamental question of density." Dellinger (2020) muses on cities after the pandemic arguing that "flattening the curve" of coronavirus by staying home "could be successfully applied at the urban scale" in relation to "smarter working" as in working from home and working more flexibly in "flattening the urban curve" for commuting and the like. Further, Dellinger (2020) sees the coronavirus pandemic as "a rare opportunity" and "a temporary break in the action of seemingly endless growth" enabling cities to "revisit what is working, what isn't, and how to adjust for the future." Madden (2020) argues that coronavirus is exposing the problems in cities associated with inequalities, providing an opportunity for change.

Gao and O'Neil (2020) provide forecasting for future city urbanization using data science and Dr. Kamran Khan, in an interview with Whitaker (2020), describes the use of computer algorithms "to detect the coronavirus outbreak" referred to as outbreak science, contributing further to the state of progress described by Houlihan and Whitworth (2019). Khan (Whitaker, 2020) explained that "understanding population movements becomes so important in anticipating how disease is spread" and to this end, how "he had spent the better part of a year persuading the airlines to share their flight data for public health" as "anonymized data." This "real time intelligence" Dr. Michael Gardam (Whitaker, 2020) explained "allowed me to stand up and say we need to pay attention to this" as in "start thinking about supplies" and "about how busy we might be." Newsom (Whitaker, 2020) noted that, "we are literally seeing in to the future and predicting in real-time based on constant update of information where patterns are starting to occur before they become headlines" suggestive perhaps of the notion of ambient approaches and inquires. Dylan George (Whitaker, 2020) called for the "need to think transformatively" urging, with other scientists, "to revive an old idea: an infectious disease forecasting center modeled on the National Weather Service." It is worth noting that Statistics Canada (2021) provides an analysis based on preliminary data to be "interpreted with caution", that "urban sprawl continues" in the Canadian cities of Toronto and Montreal "experiencing" as they are "record high population losses to surrounding areas" in the range of approximately 50,000 and 25,000, respectively, for the period covering July 1, 2019 to July 1, 2020. Statistics Canada (2021) highlights three factors contributing to the "decision of many Canadians to continue (or to no longer continue) living in large urban centres hardest hit by the pandemic" as "personal health, the ability to work remotely, and higher housing costs" while adding that "whatever the exact reasons, urban sprawl is an important trend to monitor."

3.3 Framing Ambient Approaches and Inquiries for InVisibilities in Smart Cities

While well-being has figured strongly in smart city explorations (Lin, Zhao, Yu, and Wu, 2019), strategies for health and social equity (Trencher and Karvonen, 2019), and visions for policy that take into consideration health, healthy lifestyles, and health care (Nam and Pardo, 2011), it is worth noting that McLeod, Spachos, and Plataniotis (2020) have advanced a framework for wellness assessment using the Internet of Things (IoT). In light of the coronavirus global pandemic, it seems that health needs to figure much more prominently in the design, development, and implementation of smart cities. If the range of concerns appearing in the research and popular literature is any indication, consideration will need to be given to a vast range of spaces, services, and capacities/capability from spatial arrangements to data gathering and sharing, to the design of structures and services – from public spaces, to living spaces, to working spaces, to travel (e.g., airports, airplanes, trains, buses) and other transient spaces. Anderson (2019) developed intuitive inquiry, an approach that "seeks to transform the intuitive inquirer's understanding of a research topic and to support breakthrough insights" involving what is termed "auspicious bewilderment", which she describes as the "research experience of being completely taken by surprise by insights that I did not anticipate." Intuitive inquiry could be considered for use in the context of invisibilities and visibilities in smart cities as a way possibly for people to augment their awareness, discovery, and imagination capabilities.

Additionally, smart cities and regions may have the opportunity to augment the physical by providing visible indicators and/or visualizations of the invisible, as in, making the invisible visible through urban displays, enhancing urban elements with ambient information, and the like. As the COVID-19 virus is said to be an "invisible contagion" (Paletta, 2020), it may be possible for instance for people to augment or enrich awareness through the aid of the digital infrastructure capabilities mentioned earlier by Acuto (Klaus, 2020a). Will it be possible to sense the presence of a particular virus or render such invisibilities visible with the aid of infrared or other detection and aware technologies? Indeed, the NBA (National Basketball Association) is proposing use of a "smart ring" for players to wear that will "measure body temperature, respiratory functions, and heart rate" along with being "capable of predicting COVID-19 symptoms up to three days in advance, with 90% accuracy" (Golden, 2020). Mattern (2020) considers how we sense the city in terms of sound as "urban auscultation" claiming that, "at times of crisis or change, our senses are heightened, recalibrated" adding that "*how* we listen to the city is as important as what we are listening for" speaking in terms of "tools for urban listening" that "embody particular ways of knowing the city." For Mattern (2020), "to auscultate infrastructure, to render it sensible" during this time of the pandemic "helps us appreciate how much *listening* we have ceded to machines."

In summary, by author and year, an overview of approaches and inquiries for the study of InVisibilities in smart cities that may also be amenable to the ambient are shown in Table 1. Anderson (2011) developed the Body Insight Scale (BIS) to measure well-being and other human elements through body awareness and the scale may have relevance for use in smart cities as discussed in Chapter 2. Nam and Pardo (2011) identify visions for policy that are attentive to health, healthy lifestyles, and health care. Anderson (2019) advances intuitive inquiry and the notion of auspicious bewilderment, important for discovery, insights, and the unexpected.

Jordan (2019) calls for a rethinking of artificial intelligence (AI) for human computer interactions. Lin et al. (2019) conduct explorations of well-being in smart cities, while Khan (Whitaker, 2019) articulates outbreak science using computer algorithms to detect coronavirus outbreaks. Amabile (2020) calls

Table 1. Overview of approaches and inquiries for InVisibilities in smart cities

Approaches and Inquiries for the Study of InVisibilities and the Ambient in Smart Cities		
Author(s)	**Year**	**Approaches and Inquiries**
Anderson	2011	Well-being through body awareness & the Body Insight Scale (BIS)
Nam & Pardo	2011	Policies for health and well-being in SCs
Anderson	2019	Intuitive Inquiry & auspicious bewilderment
Jordan	2019	Rethinking of AI for human computer interactions
Lin et al.	2019	Well-being explorations in SCs
Whitaker	2019	Outbreak Science – Detection and tracking software
Amabile	2020	AI Creativity – Sociotechnical systems that undermine well-being?
Chayka	2020	Architecture – Density and the future of cities
Dellinger	2020	Opportunities for flattening the curve at the urban scale – smarter working, etc.
Gao & O'Neil	2020	Defining urban land & data science for forecasting future city urbanization
Golden	2020	Smart ring to measure body well-being with predictive capabilities
Klaus	2020a/b	New imaginings for urban data and a post-pandemic future
Lytras et al.	2020	Urban computing and well-being
Madden	2020	Inequalities and opportunities for change
Mattern	2020	Listening to the city & tools for urban listening
McLeod et al.	2020	Framework for wellness assessment
Oliveira et al.	2020	Multi-tier, sensors based environmental monitoring for cyclists
Paletta	2020	Airports & associated complexities (exponential to vanishing use)
Roitblat	2020	Algorithms and AI
Schaller	2020	Density, urban edges & complexities
Ullah et al.	2020	AI, ML, and DRL applications in the evolution of smart cities
Weekes	2020	Emotional AI & well-being – Ethics, risks & social implications
Pretz	2021	AI reality check

for studies of artificial intelligence (AI) creativity in determining whether "novel socio-technical systems" will be produced that undermine human well-being. Chayka (2020) highlights the importance of density for the future of cities. Dellinger (2020) considers the notion of flattening the curve at the urban scale for smarter working and the like. Gao and O'Neil (2020) employ data science in forecasting future city urbanization while defining the notion of urban land. Golden (2020) describes a smart ring for sports players to measure body temperature and other factors while also having predictive capabilities. Acuto (Klaus, 2020a) calls for new imaginings for urban data and identifies a post pandemic future (Klaus (2020b) even as the

Figure 1. Conceptual Framework for augmented awareness for ambient InVisibilities in smart cities

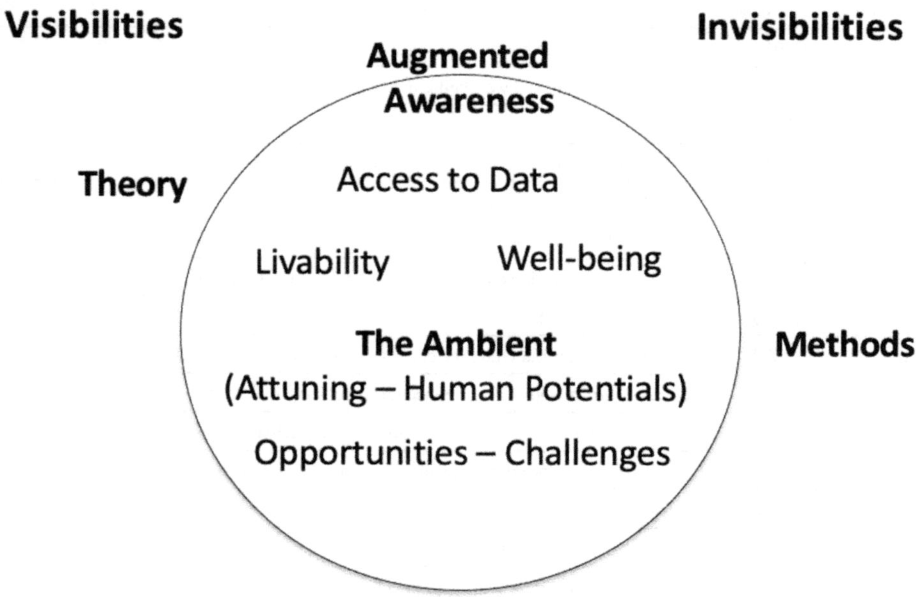

second, third, and fourth waves arrive. Madden (2020) approaches challenges such as those surfaced by the global COVID-19 pandemic as opportunities for change. Mattern (2020) draws attention to "urban auscultation" and the importance of how we are listening to the city. McLeod et al. (2020) utilize the Internet of Things (IoT) as a framework for wellness assessment while Oliveira et al. (2020) assess the quality of bike paths using a sensors-based environmental monitoring approach.

Paletta (2020) articulates the complexities brought about by the global COVID-19 pandemic using the example of airports and the movement from exponential growth in usage to abrupt and vanishing use, Roitblat (2020) articulates the limitations of algorithms for artificial intelligence, while Schaller (2020) highlights the challenges and complexities of urban density and creativity. The role of artificial intelligence (AI), machine learning (ML), and deep reinforcement learning (DML) applications in the evolution of smart cities is reviewed by Ullah et al. (2020) and Weekes (2020) explores the ethics, risks, and social implications of emotional AI as an approach to well-being. Jordan encourages a reality check for artificial intelligence (AI) from the perspective of machine learning (ML) (Pretz, 2021) and the need for engineering to incorporate considerations of human happiness in technology development.

Based on the review of the research literature presented in this chapter, a conceptual framework for augmented awareness for ambient InVisibilities in smart cities is advanced as depicted in Fig 1.

Augmented awareness involves opening to increased capacities for attuning to urban environments and regions and human potentials more generally, in relation to the dimensions of visibilities and invisibilities in smart cities, encompassing data (e.g., access to and sharing of) whether public or personal; livability (e.g., air quality, etc.); and overall well-being, supported by theory and methods and approaches, and where challenges give way to opportunities.

Through emerging understandings of visibilities and invisibilities as in urban areas and regions, relationships are explored in this chapter in responding to the research question posed in the introduction of this chapter, and restated as a proposition under exploration in this chapter, as follows:

P: Understanding the nature of the relationship(s) between the ambient and InVisibilities as well-being is important in smart cities and regions in relation to emergent and evolving spaces for humanity and more specifically for aware people interacting with each other and with aware technologies.

4. WELL-BEING AND ATTUNING: AMBIENT INQUIRY IN SMART CITIES

Based on guidance from the literature review in this chapter the research design for this chapter uses an exploratory case study approach, said (Yin, 2018) to be particularly appropriate for the study of contemporary phenomena, which in this chapter focuses on *air quality* as a proxy for *livability* and *attuning to urban spaces* as a proxy for *the ambient* in smart cities and regions. An explanatory correlational design is also incorporated into the research design in this chapter, because according to Yin (2018) this provides "complementarity of case study and statistical research." The study upon which this chapter is based was described using a website where, during sign up, demographic data were collected including elements such as age range, location (city), and people could self-identify in one or more categories (e.g., community member, city official, researcher, educator, student, business, etc.). Upon sign up, people were invited to complete a survey about their experience of their city as a smart city. An invitation was also provided to participate in an in-depth interview about smart cities, online or in person, depending upon location. The study attracted participation from across Canada and extended to other cities in the United States, Ireland, Greece, and Israel. The survey instrument, containing a mix of closed and open-ended questions, and the interview protocol were pretested prior to use in the study. To enable responses from a broad spectrum of people, heterogeneity sampling, a type of purposive sampling was used whereby an invitation to participate in the study was posted to webspaces with the potential to attract a wide response. Data analysis included the use of content analysis for qualitative data and descriptive statistics for quantitative data. The Real Statistics Software Resource Pack Add In (Zaiontz, 2019) for Microsoft Excel was used, consisting as it does of a wide range of tools for statistical analysis, such as the correlation feature for calculation of the Spearman correlation for the ordinal data, generated by assessments from survey respondents. The survey instrument and interview protocol were designed to enable people to assess their experience of the city and region where they live in a wide variety of ways including aspects of well-being and livability in urban spaces. The data collected in parallel with this study through group and individual discussions included a wide range of people across many sectors in small to medium to large-sized cities in Canada (e.g., greater Victoria, Vancouver, and Toronto). Overall, data were analyzed for n=78 including 41% females and 59% males for people ranging in age from their 20s to their 70s.

4.1 Attuning and InVisibilities in Smart Cities

Survey respondents were asked to assess to what extent *air quality* contributes to the *livability* of a smart city (on a scale of 1 to 7 where 1 = not at all and 7 = absolutely). As indicated in Table 2, 25% responded at the neutral position 4 on the 7-point Likert-type scale and 75% at the upper end of the scale at

Table 2. Assessments for well-being (air quality) and the Ambient (attuning to urban spaces)

Items	Assessments	Correlation (Spearman)
Air Quality (livability/well-being)	25% (4); 75% (7)	.80
Attuning to Urban Spaces (the Ambient)	17% (4); 33% (6); 50% (7)	

position 7. When asked to assess to what extent *attuning to urban spaces* through city-focused social media and other aware technologies give rise to many possibilities in smart cities, 17% responded at position 4 on the scale, 33% at position 6, and 50% at position 7.

The correlation between *livability* as *well-being* using *air quality* as a proxy and *the ambient* using the proxy of *attuning to urban spaces* as ordinal data, was conducted using a feature of the Real Statistics Software (Zaiontz, 2019) in Microsoft Excel showing the Spearman correlation coefficient result of .80 in Table 2.

Probing further, survey respondents were asked to assess to what extent *opportunities to easily choose whether and when to share non-sensitive data* contributes to the *livability* of a smart city (on a scale of 1 to 7 where 1 = not at all and 7 = absolutely). Table 3 shows that, 25% responded at the neutral position 4 on the 7-point Likert-type scale, 25% at position 5, and 50% at the upper end of the scale at position 7. The correlation between *livability* as *well-being* using *share non-sensitive data* as a proxy and *the ambient* using the proxy of *attuning to urban spaces* as ordinal data, shows a positive Spearman correlation coefficient result of .48 in Table 3.

Probing the data variable further, survey respondents were asked to assess to what extent *access to public data* contributes to the *livability* of a smart city (on a scale of 1 to 7 where 1 = not at all and 7 = absolutely). As indicated in Table 4, 25% responded at the neutral position 4 on the 7-point Likert-type scale; 25% at position 5, 25% at position 6, and 25% at the upper end of the scale at position 7.

Table 3. Assessments for sharing non-sensitive data and the Ambient (attuning to urban spaces)

Items	Assessments	Correlation (Spearman)
Share Non-Sensitive Data (livability)	25% (4). 25% (5); 50% (7)	.48
Attuning to Urban Spaces (the Ambient)	17% (4); 33% (6); 50% (7)	

Table 4. Assessments for access to public data and the Ambient (attuning to urban spaces)

Items	Assessments	Correlation (Spearman)
Access to Public Data (livability)	25% (4); 25% (5); 25% (6); 25% (7)	.62
Attuning to Urban Spaces (the Ambient)	17% (4); 33% (6); 50% (7)	

The correlation between *livability* using *access to public data* as a proxy and *the ambient* using the proxy of *attuning to urban spaces*, shows a positive Spearman correlation coefficient result for ordinal data of .62 in Table 4.

Qualitatively, in describing the notion of *attuning to urban spaces* in terms of the pulse of the city, an individual in business identified "the multi layers of senses, and the feeling of dynamic." It is also worth noting that a public administrator indicated that feeling the pulse of the city is enabled by "events of a sociopolitical and cultural nature."

In terms of *well-being* and what constitutes evolving notions of everyday *livability*, as cities transform before our eyes during the global coronavirus pandemic, an educator observed that, "office blocks may steadily (or even quickly) become a thing of the past as post pandemic" or what turns out to be spaces between pandemic waves, "we are discovering that many office jobs can be done from home."

4.2 Discussion and Interpretation of Findings

The correlation found in this work between *well-being* using *air quality* as a proxy (as air quality may be both visible or invisible and as such, an indicator of InVisibilities) for *livability* and *the ambient* using *attuning to urban spaces* as a proxy yields a positive correlation of .80. According to Creswell (2018), a correlation between .66 and .85 indicates a "very good" relationship and "good prediction can result from one variable to the other." Also related to well-being, it is worth noting that, an educator noticed the impact of pandemic lockdowns on office space and the shift to working from home as a possibly longer-term practice.

As such, findings in this chapter may be promising for ambient inquiries and approaches associated with *attuning to urban spaces* in evolving understandings of InVisibilities in the form of *livability* factors such as air quality, working spaces, and so on in smart cities. Smart cities and regions are currently faced with high levels of uncertainty in relation to the current global coronavirus pandemic, pointing to opportunities for refining understandings of *well-being* and *livability*. For example, urban areas and regions are undergoing rapid transformations associated with the global pandemic and the uncertainties associated with such transformations could affect the sense of well-being or livability people are experiencing, making access to data and information all the more critical. Also of note is the indication that "events of a sociopolitical and cultural nature" enable people to feel the pulse of the city, possibly of value for events in many cities that are occurring in virtual spaces.

5. EMERGING RESEARCH AND OPPORTUNITIES

Emerging trends are identified in this chapter for practice and research in sections 5.1 and 5.2 pertaining to the viability of the framework advanced for augmented awareness for ambient InVisibilities in smart cities.

5.1 Practice

In practical terms, this chapter contributes to explorations of well-being as a way of rethinking methods and approaches to addressing challenges associated with livability and what is made visible or invisible in smart cities and to the surfacing of associated variables and indicators.

Ambient Inquiries. This chapter provides examples of the use of well-being to drive inquiries about smart city visibilities and invisibilities as Invisibilities associated with livability and attuning to urban spaces as ambient. Crucially, this work invites participation involving approaches to data access and sharing for to urban problem-solving that are more dynamic, adaptive, and responsive to contemporary urban challenges and opportunities, that take into consideration the needs and environments of people.

Variables and Indicators. While this chapter surfaces the well-being variable of air quality as potentially encompassing both the invisible and the visible, hence *InVisibilities* in relation to variables such as the ambient using attuning to urban spaces, associated as it is with urban design, the potential for learning more about these variables exists as well as opportunities for identifying and working with many other possible variables.

5.2 Research

In terms of future research opportunities and emerging discourse spaces, this chapter advances the potential for urban methods, approaches, and inquiries as ambient in the context of visibilities and invisibilities as In/Visibilities in smart cities.

Ambient Methodologies

The viability of the conceptual framework for augmented awareness for ambient invisibilities in smart cities operationalized for use in this chapter demonstrates early-stage potentials while offering and opening up new understandings for evolving urban spaces through further testing and validation.

Ambient InVisibilities and Well-Being

Conceptualization of spaces for real-time, dynamic, and adaptive types of approaches to making the invisible visible emerge in this chapter through explorations of well-being in terms of livability, opening the way for new techniques and possibly novel roles and collaborations for citizens, businesses, researchers, and governments involving the ambient and attuning to urban spaces. As such, areas for continued exploration potential include:

1. *Air Quality* explored through urban initiatives focusing on *attuning to urban spaces* and well-being in public spaces taking people and their experiences into consideration and possibly utilizing dynamic and interactive assessments to learn more about what contributes to health, well-being, and livability concerns in urban environments and regions
2. *Sharing of Non-Sensitive Data* based on ease in choosing whether and when, explored in relation to livability opens spaces for people to think about data, different types of data, and data options in smart environments
3. *Access to Public Data* explored in relation to livability opens spaces for people to think about the notion of public data and the relevance of such data for everyday life

Ambient Approaches

This chapter opens the way to future research opportunities and challenges related to the potential for developing approaches for attuning to urban spaces, as a proxy for the ambient, with implications for practice.

Ambient Inquiry

This chapter opens the way to future research opportunities and challenges related to the potential for movement toward the building of ambient inquiry as an urban research technique involving the use of well-being and other indicators of livability, in real-time, with implications for practice.

Additionally, this chapter directs attention to the potential for augmenting awareness through the use of real-time urban data.

6. CONCLUSION

This chapter provides an exploration and assessment of the use of *livability* to represent urban InVisibilities using the proxy of *air quality* in relation to *attuning to urban spaces* to represent *the ambient*, in articulating the notion of ambient approaches and inquires for smart cities and regions. The *sharing of non-sensitive data* and *access to public data* are also explored as experienced by people in their city in relation to *attuning to urban spaces*. Responses are then correlated, resulting in what is considered to be a "very good" relationship (Creswell, 2018) between air quality and attuning to urban spaces based on a Spearman correlation coefficient of .80 while the correlation between access to public data and attuning to urban spaces is said to be "useful for limited prediction" (Creswell, 2018), based on a Spearman correlation coefficient of .62. The main contributions of this chapter include a) articulation and operationalization of a conceptual framework for augmented awareness for ambient InVisibilities in smart cities; and b) advancing ambient approaches and inquires that take into consideration the intangible and visibilities and invisibilities as InVisibilities in relation to emerging technologies while being attentive to attuning to and by people, thus augmenting awareness in fostering smarter, more responsive, cities and regions.

Possible limitations of this chapter relate to the small sample size underlying the study for this work as well as the use of proxies for *livability* (*air quality*) in relation to *the ambient* (*attuning to urban spaces*). Further limitations related to literature review domain coverage are mitigated by the interdisciplinary potential to extend this review to other areas going forward. Among other limitations of this work is that of city size as well as geographic location and this is mitigated by the potential to extend this work going forward to respondents in other much larger cities, megacities, and countries.

Findings in this chapter advance variables relevant to *InVisibilities* and more ambient approaches to the study, practice, and evolving of smart cities while possibly extending what McKenna (2019) refers to as ambient urbanities.

Insights of particular interest to the reader include the following:

1. The potential for opportunities to assess *livability* in terms of *air quality* as understandings of InVisibilities in smart cities
2. The importance of assessments of *sharing non-sensitive data* in learning more about InVisibilities in urban environments

3. The surfacing of variables associated with *access to public data* as *InVisibilities* in relation to *attuning to urban spaces* as *the ambient* for informing approaches, methods, and inquiries for urban research and practice
4. The presence of a range of relationships between *data* and *the ambient* (*attuning to urban spaces*) in smart cities

Ideas highlighted in this chapter pertain to the following:

1. The importance of assessments of *livability* as a way for people to become more involved in the invisible and intangible aspects of smart cities phenomena
2. The potential for people to shed more light on invisible spaces in digital/physical urban spaces through assessments of their experiences of sharing non-sensitive data and accessing public data in their city
3. Ambient inquiries and methods as novel approaches to the study and practice of smarter cities involving more aware and attuned people in urban environments and regions

Exercises for the reader to consider, encouraging further exploration include the following:

1. Identify one or more examples of *well-being* and *attuning to urban spaces* in your city
2. Identify one or more examples of *livability* and *attuning to urban spaces* in your city
3. Describe why *sharing non-sensitive data* and *attuning to urban spaces* could be important for smart cities, based on your example(s)
4. Describe why *access to public data* and *attuning to urban spaces* could be important for smart cities, based on your example(s)

If you would like to share your responses to the Exercises with the author, please complete the questions in this online space for Chapter 7 – Ambient Approaches and Inquiries

This work will be of interest to a broad audience including students, educators, researchers, artists, urban planners, urban practitioners, designers, developers, policymakers, public health officials, community members, government officials, place-makers, and anyone concerned with livability, health and wellbeing, and emerging ambient approaches, methodologies, and inquiries that take into consideration the complexities, uncertainties, and transformations contributing to urban data challenges and opportunities posed in renderings of InVisibilities (e.g., the invisible and visible) in smart, responsive, and future cities.

REFERENCES

Acuto, M. (2020). COVID-19: Lessons for an urban(izing) world – Commentary. *One Earth, 2*(4), 317-319. doi:10.1016/j.oneear.2020.04.004

Amabile, T. M. (2020). Guidepost: Creativity, artificial intelligence, and a world of surprises. *Academy of Management Discoveries*, *6*(3), 351–354. doi:10.5465/amd.2018.0075

Amabile, T. M., & Hall, D. T. (2019). The undervalued power of self-relevant research: The case of researching retirement while retiring. *Academy of Management Journal*. Advance online publication. doi:10.5465/amp.2018.0083

Anderson, R. (2011). Body insight scale. *Mind Garden Blog*. Retrieved 31 May 2020 from https://www.mindgarden.com/73-body-insight-scale

Anderson, R. (2019). Intuitive inquiry: Inviting transformation and breakthrough insights in qualitative research. *Qualitative Psychology*, *6*(3), 312–319. doi:10.1037/qup0000144

Berman, M. (1982). *All that is solid melts into air: The experience of modernity*. Simon & Schuster.

Chayka, K. (2020). How the coronavirus will reshape architecture: What kinds of space are we willing to live and work in now? *The New Yorker*. Retrieved 19 June 2020 from https://www.newyorker.com/culture/dept-of-design/how-the-coronavirus-will-reshape-architecture

Connolly, C., Keil, R., & Ali, S. H. (2020). Extended urbanisation and the spatialities of infectious disease: Demographic change, infrastructure and governance. *Urban Studies (Edinburgh, Scotland)*, 1–19. doi:10.1177/0042098020910873

Creswell, J. W. (2018). *Educational Research: Planning, Conducting, and Evaluating Quantitative and Qualitative Research* (6th ed.). Pearson.

Dellinger, A. J. (2020). What will cities look like after coronavirus? *Connected Cities Lab, News, Mic*. Retrieved 23 May 2020 from https://www.mic.com/p/what-will-cities-look-like-after-coronavirus-22892442?utm_campaign=mic&utm_content=1589397522&utm_medium=owned&utm_source=facebook,twitter

Gao, J., & O'Neil, B. C. (2020). Mapping urban global land for the 21st century with data-driven simulations and shared socioeconomic pathways. *Nature Communications*, *11*(2302), 2302. Advance online publication. doi:10.103841467-020-15788-7 PMID:32385275

Golden, J. (2020). Inside the NBA's plan to use smart technology and big data to keep players safe from coronavirus. *CNBC Sports*. Retrieved 19 June 2020 from https://www.cnbc.com/2020/06/17/nba-coronavirus-plan-smart-technology-to-keep-players-safe.html

Houlihan, C. F., & Whitworth, J. A. (2019). Outbreak science: Recent progress in the detection and response to outbreaks of infectious diseases. *Clinical Medicine (London, England)*, *19*(2), 140–144. doi:10.7861/clinmedicine.19-2-140 PMID:30872298

Hsiang, S., Allen, D., Annan-Phan, S., Bell, K., Bolliger, I., Chong, T., Druckenmiller, H., Huang, L. Y., Hultgren, A., Krasovich, E., Lau, P., Lee, J., Rolf, E., Tseng, J., & Wu, T. (2020). The effect of large-scale anti-contagion policies on the COVID-19 pandemic. *Nature*, *584*(7820), 262–267. Advance online publication. doi:10.103841586-020-2404-8 PMID:32512578

Jordan, M. I. (2019). Artificial intelligence—The revolution hasn't happened yet. *Harvard Data Science Review*, *1*(1). Advance online publication. doi:10.1162/99608f92.f06c6e61

Keil, R. (2020). The limits of global urbanization and the challenges to planning. *disP - The Planning Review, 56*(1), 4-7. doi:10.1080/02513625.2020.1756620

Klaus, I. (2020a). Pandemics are also an urban planning problem. *CityLab*. Retrieved 23 May 2020 from https://www.citylab.com/design/2020/03/coronavirus-urban-planning-global-cities-infectious-disease/607603/

Klaus, I. (2020b). The post-pandemic urban future is already here. *CityLab*. Retrieved 23 May 2020 from https://www.citylab.com/design/2020/04/coronavirus-urban-planning-cities-architecture-history/609262/

Lin, C., Zhao, G., Yu, C., & Wu, Y. J. (2019). Smart city development and residents' well-being. *Sustainability, 11*(3), 676. doi:10.3390u11030676

Lytras, M. D., Visvizi, A., Torres-Ruiz, M., Damiani, E., & Jin, P. (2020). IEEE Access Special Section Editorial: Urban computing and well-being in smart cities: Services, applications, policymaking considerations. *IEEE Access: Practical Innovations, Open Solutions, 8*, 72340–72346. doi:10.1109/ACCESS.2020.2988125

Madden, D. (2020). Our cities only serve the wealthy: Coronavirus could change that. *The Guardian*. Retrieved 2 June 2020 from https://www.theguardian.com/commentisfree/2020/jun/02/coronavirus-cities-elite-community

Mattern, S. (2020). Urban auscultation; or, perceiving the action of the heart: How we listen to the city is as important as what we are listening for. *Places Journal*. Retrieved 22 June 2020 from https://placesjournal.org/article/urban-auscultation-or-perceiving-the-action-of-the-heart/

McCann, E. (2020). Listening to the city in a global pandemic. In What's it like to live in a city as a global pandemic takes hold? *CityRoadPodcast*. Retrieved 23 May 2020 from https://cityroadpod.org/2020/03/29/listening-to-the-city-in-a-global-pandemic/

McKenna, H. P. (2019). Visibilities and invisibilities: Methodologies and theoretical spaces for ambient urbanities. In *Ambient Urbanities as the Intersection between the IoT and the IoP in Smart Cities* (pp. 192–222). IGI Global. doi:10.4018/978-1-5225-7882-6.ch007

McLeod, K., Spachos, P., & Plataniotis, K. (2020). An experimental framework for wellness assessment through the Internet of Things. *IEEE Internet Computing, 24*(2), 8–17. Advance online publication. doi:10.1109/MIC.2020.2972817

Nam, T., & Pardo, T. A. (2011). Smart city as urban innovation: Focusing on management, policy, and context. *ICEGOV2011*, 185-194.

Oliveira, F., Costa, D. G., & Dias, A. (2020). A multi-tier sensors-based environmental monitoring approach to assess the quality of bike paths in urban areas. In *2020 IEEE International Smart Cities Conference (ISC2)* (pp. 1-4). IEEE. 10.1109/ISC251055.2020.9239071

Paletta, A. (2020). Why airports die: Expensive to build, hard to adapt to other uses, and now facing massive pandemic-related challenges, airport terminals often live short, difficult lives. *CityLab*. Retrieved 22 May 2020 from https://www.citylab.com/design/2020/05/air-travel-coronavirus-airport-design-historic-preservation/611326/

Pretz, K. (2021). Stop calling everything AI, machine learning pioneer says. *IEEE Spectrum*, *31*(March).

Roitblat, H. L. (2020). *Algorithms Are Not Enough: Creating General Artificial Intelligence*. MIT Press. doi:10.7551/mitpress/11659.001.0001

Schaller, B. (2020). Density isn't easy. But it's necessary. *CityLab*. Retrieved 23 May 2020 from https://www.citylab.com/perspective/2020/05/coronavirus-urban-density-history-traffic-congestion-disease/611095/

Statistics Canada. (2021). *Canada's population estimates: Subprovincial areas, July 1, 2020*. Released 14 January 2021. Retrieved 1 May 2021 from https://www150.statcan.gc.ca/n1/daily-quotidien/210114/dq210114a-eng.htm?HPA=1

Trencher, G., & Karvonen, A. (2019). Stretching "smart": Advancing health and well-being through the smart city agenda. *Local Environment*, *24*(7), 610–627. doi:10.1080/13549839.2017.1360264

Ullah, Z., Al-Turjman, F., Mostarda, L., & Gagliardi, R. (2020). Applications of artificial intelligence and machine learning in smart cities. *Computer Communications*, *154*, 313–323. doi:10.1016/j.comcom.2020.02.069

Watts, J. (2020). Interview – Bruno Latour: 'This is a global catastrophe that has come from within'. *The Guardian*. Retrieved 6 June 2020 from https://www.theguardian.com/world/2020/jun/06/bruno-latour-coronavirus-gaia-hypothesis-climate-crisis

Weeden, S. A. (2020). Do you have a right to go to the cottage during the coronavirus pandemic? *The Conversation*. Retrieved 22 May 2020 from https://theconversation.com/do-you-have-a-right-to-go-to-the-cottage-during-the-coronavirus-pandemic-138702

Weekes, S. (2020). UK-Japan project to examine use of emotional AI in smart cities. *Smart Cities World: Sharing Ideas to Solve Urban Challenges*. Retrieved 4 June 2020 from https://www.smartcitiesworld.net/news/news/uk-japan-project-to-examine-use-of-emotional-ai-in-smart-cities-4974

Whitaker, B. (2020). The computer algorithm that was among the first to detect the coronavirus outbreak. *CBS News, 60 Minutes*. Retrieved 28 April 2020 from https://www.cbsnews.com/news/coronavirus-outbreak-computer-algorithm-artificial-intelligence/

Zaiontz, C. (2019). *Real Statistics using Excel*. www.real-statistics.com

ADDITIONAL READING

Bliss, L. (2020). Elevators changed cities. Will coronavirus change elevators? CityLab. Retrieved 23 May 2020 from https://www.citylab.com/transportation/2020/05/elevator-anxiety-coronavirus-health-risks-design-history/611032/

Kling, S. (2020). Is the city itself the problem? *CityLab*. Retrieved 23 May 2020 from https://www.citylab.com/perspective/2020/04/coronavirus-cases-urban-density-suburbs-health-parks-cities/610210/

Muggah, R., & Florida, R. (2020). Megacity slums are incubators of disease – but coronavirus response isn't helping the billion people who live in them. *The Conversation*. Retrieved 17 May 2020 from https://theconversation.com/megacity-slums-are-incubators-of-disease-but-coronavirus-response-isnt-helping-the-billion-people-who-live-in-them-138092

Poon, L. (2020). A lesson from social distancing: Build better balconies. CityLab. Retrieved 23 May 2020 from https://www.citylab.com/life/2020/04/apartment-design-balcony-private-outdoor-space-zoning-laws/610162/

Roberts, K. B. (2020). Forecasting urbanization: Data science drives new maps to predict the growth of cities over next century. *UDaily*. University of Deleware. Retrieved 20 May 2020 from https://www.udel.edu/udaily/2020/may/mapping-global-urban-expansion-21st-century/

Rosalsky, G. (2020). What is the future of cities? *PlanetMoney: The Economy Explained – NPR*. Retrieved 20 May 2020 from https://www.npr.org/sections/money/2020/05/19/858068115/what-is-the-future-of-cities

Safi, M. (2020). Coronavirus will reshape our cities – we just don't know how yet: The development of cities has been affected by disease for centuries, so what legacy will Covid-19 leave on urban life? *The Guardian*. Retrieved 22 May 2020 from https://www.theguardian.com/world/2020/may/22/coronavirus-will-reshape-our-cities-we-just-dont-know-how-yet

Surico, J. (2020). Need more outdoor public space? Maybe cities already have it: With parks filled and social distancing in effect, cities need to find more room for residents to get outside during lockdowns. Here's where it's hiding. *CityLab*. Retrieved 23 May 2020 from https://www.citylab.com/design/2020/05/parks-outdoor-public-areas-golf-course-cemetery-parking-lot/611257/

KEY TERMS AND DEFINITIONS

Ambient Inquiries: Ambient inquiries refer to approaches to research and practice that are more adaptive and flexible, involving increased awareness of and attuning to context and environment.

Ambient Methodologies: Ambient methodologies refer to more adaptive approaches to explorations of awareness in relation to people and awareness-enabled and enabling technologies encompassing the physical, digital, and less tangible, but not limited to ambiences, atmospheres, augmentations, emotion/affect, and the like.

Ambient Theory: Ambient theory refers to an emerging new addition to existing theories for urban research that is more amenable to adaptive frameworks for smart, responsive, and future cities while enabling possibly more aware, dynamic, interactive, and evolving models in response to increasing complexities and uncertainties.

Ambient Visibilities/Invisibilities: Ambient visibilities/invisibilities refer to urban elements accommodating more aware people and awareness-enabled technologies encompassing the physical, the digital, and the less tangible.

Augmentations: Augmentations refer to the extending of capabilities involving the use of technologies, as in for example, augmented reality and the visual overlay of information in everyday spaces, contributing to heightened forms of awareness.

Awareness: Awareness is conceptualized in this work as the quality of being aware as it applies to people on the one hand, to the enabling of technologies on the other, and to the enhancing of awareness through people interacting with awareness-enabled and enabling technologies.

Invisibilities: Invisibilities refer to urban elements that are rendered invisible or that may present as intangibles.

Visibilities: Visibilities refer to physical and other visible urban elements and infrastructures.

Chapter 8
A Synthesis and Taxonomy:
Visibilities and Invisibilities in Smart Cities

ABSTRACT

The purpose of this chapter is to provide a synthesis of urban visibilities and invisibilities in smart cities and regions as presented throughout this book. Key elements including dimensions, relationships, theory, and methods are presented in developing an integrated conceptual framework for the interweaving of the visible and invisible as InVisibilities in relation to the ambient in smart cities. Tables are included that enable summary and comparative views across chapters pertaining to findings, insights, ideas, exercise questions, and future directions for research and practice. This chapter makes a contribution to 1) emerging research and practice for ambient InVisibilities in smart cities; 2) the synthesizing of urban theory and methods pertaining to the visible and invisible dimensions of 21st century cities and urban regions; and 3) the extending of understandings pertaining to smart cities through of a conceptual framework for visibilities, invisibilities, and the ambient encompassing the physical and digital as a guide for theory and practice going forward.

DOI: 10.4018/978-1-7998-3850-0.ch008

1. INTRODUCTION

This chapter provides a summary and synthesis of emerging understandings of visibilities and invisibilities as fundamental elements of 21st century cities and regions. Dimensions of visibilities and invisibilities emerge across transparency and translucence in relation to the ambient in the context of smart cities and regions. The nature of theory and methods in support of the study and practice of working with visibilities and invisibilities takes form in this work and opportunities arise going forward for research and practice in smart cities and regions. A key contribution of this chapter is the formulation of a conceptual framework for visibilities, invisibilities, and the ambient in smart cities as an integration of frameworks presented in earlier chapters of the book. This chapter is significant in that it provides a taxonomy of visibilities and invisibilities in smart cities important for the advancing of ambient theory and ambient inquiries in support of the evolving of urban theory and methods.

Objective: The objective of this chapter is to integrate, summarize, and advance evolving perspectives on, and approaches to, understandings of visibilities and invisibilities in smart cities and regions through the presentation of images and tables, as follows: a conceptual framework, observations from the literature reviews, research questions and propositions explored, correlations between variables, insights, ideas, exercise questions, future directions as opportunities for research and practice, and a taxonomy of visibilities and invisibilities for smart cities.

1.1 Background

The notion of ambient InVisibilities is not new in that the concept was advanced in the research literature (McKenna, 2019) along with a framework "organized around aware people and aware technologies, encompassing the physical and digital in relation to ambiences, atmospheres, and augmentations." What is new in this work is that additional explorations are provided contributing to a rich and evolving interdisciplinary perspective. Additionally, the theoretical perspective and methods and approaches offered in support of research and practice are presented in the form of a conceptual framework and a taxonomy for visibilities, invisibilities and the ambient in smart cities and regions.

1.2 Overview

A summary and synthesis of the interwoven nature of visibilities and invisibilities as InVisibilities is presented in this chapter in the context of smarter cities involving more aware people interacting with more aware technologies, contributing to contemporary renderings of the ambient.

Definitions are presented here, based on the research literature, for key terms used throughout this work. Additional terms with definitions, as articulated by this chapter, are included in the *Key Terms and Definitions* section at the end of this chapter.

Ambient InVisibilities: According to McKenna (2019), ambient InVisibilities derive from a combination of visibilities and invisibilities and pertain to "urban elements accommodating more aware people and aware technologies encompassing the physical and digital in relation to ambiences, atmospheres, and augmentations contributing further to the ambient turn."

As with the other chapters in this book, the primary audiences for this chapter include city officials, community members, students, educators, researchers, policy makers, urban practitioners, city managers, urban planners, urban designers, businesses, engineers, architects, public administrators, and anyone concerned with opportunities for leveraging the potentials of visibilities and invisibilities for more aware people in smarter, more sustainable, and livable cities.

2. VISIBILITIES, INVISIBILITIES AND THE AMBIENT: PUTTING IT ALL TOGETHER

This section provides an overview of key elements constituting the interweaving of visibilities and invisibilities as InVisibilities, in smarter more aware environments in support of the ambient, beginning with the various conceptual frameworks, followed by variables and their relationships, insights, ideas, exercises, and future directions as opportunities for research and practice.

2.1 Integrating Invisibilities, Visibilities, and the Ambient Into a Framework

Bringing together the conceptual frameworks developed and operationalized in chapters one to seven of this book enables development of an integrated conceptual framework for visibilities, invisibilities, and the ambient in smart cities and regions as presented in Fig. 1. Chapter one identified and described dimensions of visibilities and invisibilities in smart cities in relation to physical, digital, and human infrastructures. *ICTs* (information and communication technologies) as a proxy for invisible infrastructures were explored and correlated with visible infrastructures. Visible and invisible infrastructures were also explored in relation to people and their experience of attuning to urban spaces through feeling the pulse of the city. Chapter two identified and described approaches to making the invisible visible through exploring and correlating *heightening urban sensibilities* with *urbanizing* (e.g., adapting for urban uses). Emotion/affect was also explored through feelings of body tension in the city, in the form of anger in relation to awareness. Chapter three identified and described the dimension of transparency for visibilities and invisibilities in smart cities in relation to the ambient and this was explored using *openness* as a proxy for transparency and *attuning to urban spaces* as a proxy for the ambient. Chapter four identified and described the dimension of translucence for visibilities and invisibilities in smart cities in relation to the ambient and this was explored using *visible infrastructures* as a proxy for transparency and *awareness* as a proxy for the ambient.

Chapter five identified and described theory for visibilities and invisibilities in smart cities in relation to the ambient and this was explored using *ICTs* as a proxy for *visibilities* and *invisibilities* as *InVisibilities* and *awareness* as a proxy for *the ambient*.

Chapter six identified and described methodologies for *visibilities* and *invisibilities* in smart cities in relation to *the ambient* and this was explored using *placemaking* as a proxy for *invisibilities and visibilities* as InVisibilities and *attuning to urban spaces* as a proxy for *the ambient*.

Chapter seven explored well-being in terms of *livability* using *air quality* as a proxy in relation to *attuning to urban spaces* as a proxy for *the ambient, contributing to understandings of augmented awareness*. Proxies for livability were also used pertaining to data including *opportunities to choose whether and with whom to share your non-sensitive data* and *access to public data* in relation to *the ambient*. InVisibilities could be extended to other approaches discussed in the chapter such as design thinking, data visualizations, self-relevant research (Amabile and Hall, 2019), to name a few. It should be noted that going forward, explorations and correlations for any given dimension of this framework could be extended beyond those used in this book. Indeed, it is expected that additional dimensions for visibilities, invisibilities, and the

Figure 1. Conceptual framework for visibilities, invisibilities and the ambient in smart cities

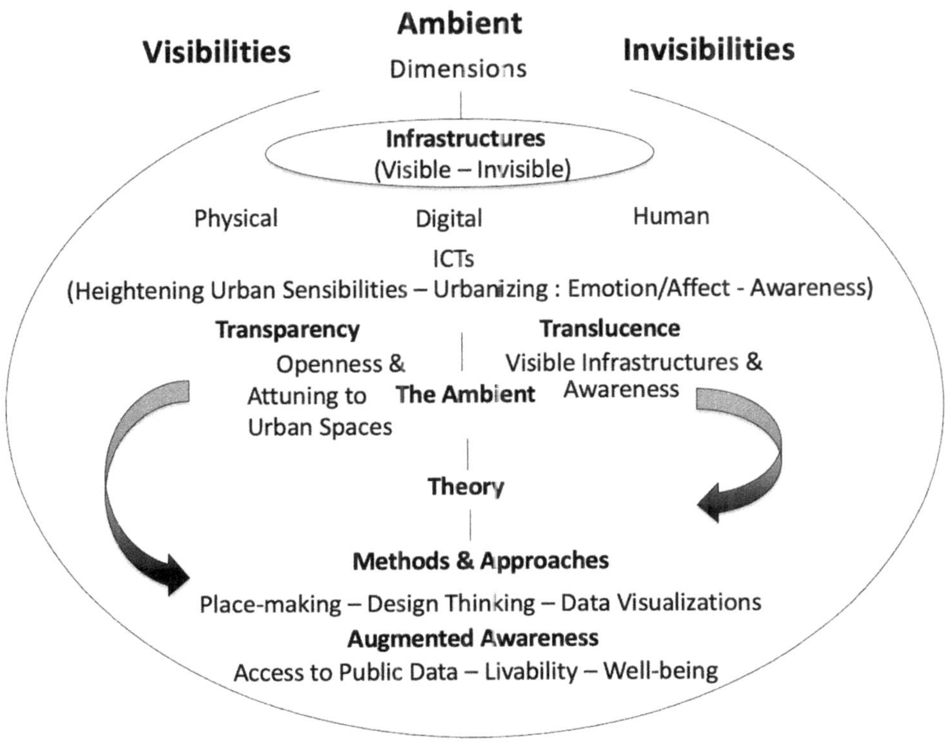

ambient in smart cities and regions will be identified and described going forward.

2.2 Observations From the Literature Reviews

Table 1 brings together observations emerging from the literature reviews for visibilities and invisibilities in smart cities corresponding to the focus and dimensions for each of the seven chapters in the book. For Chapter 1, perspectives on smart cities, intersections emerge for information and measurements, the virtual in terms of the digital, technology and data in terms of algorithms and unnoticed data collection, augmented reality and the sensorial. For Chapter 2, making the invisible visible, approaches that emerged include art, computational intelligence (CI), communication networks (CN), information, technology, data, mapping, visualization and infrastructures, human augmentation, and sensors for sensing.

Table 1. Observations from the literature reviews for visibilities and invisibilities in smart cities

Focus and Dimensions	Observations
1: Perspectives on Smart Cities *Visible Infrastructures* *Invisible Infrastructures (ICTs)*	• Intersections emerge for information & measurements, the virtual (e.g., virtual expanse & the digital), technology & data (e.g., algorithms & unnoticed data collection), augmented reality & the sensorial
2: Making the Invisible Visible *Heightening Urban Sensibilities* *Urbanizing*	• Approaches: art, CICN, information, technology, data, mapping, visualization, infrastructures (visible/invisible), human augmentation & sensors for sensing
3: Transparency & The Ambient *Openness* *Attuning to Urban Spaces*	• Intersections emerge for data. Openness appears in relation to transparency and the presence of AI, AmI & multiexperience.
4: Translucence & The Ambient *Visible Infrastructures* *Awareness*	• Intersections emerge for the social & awareness in relation to translucence • Experiences emerge for the ambient as well as the presence of complexity, uncertainty & sensors
5: Theory in Smart Cities *ICTs* *Awareness*	• Across elements characterizing InVisibilities & theory in smart cities, ambience, change & appreciative inquiry, AmI & complexity
6: Methods & Approaches *Placemaking* *Attuning to Urban Spaces*	• Context & imagination appear • Importance of people and the presence of emotion, attuned & ambiance
7: Ambient Approaches & Inquiries *Livability/Well-being (Air Quality)* *Attuning to Urban Spaces (Ambient)*	• Augmenting of well-being and livability awareness with data • Implications for access to public data in real time & health monitoring on smart devices

For Chapter 3, transparency and the ambient, across characteristics intersections emerge for data while openness appears in relation to transparency and the presence of artificial intelligence (AI), ambient intelligence (AmI), and multiexperience. For Chapter 4, translucence and the ambient, intersections emerge for the social and awareness in relation to translucence while experiences emerge for the ambient, as well as the presence of complexity, uncertainty, and sensors. For Chapter 5, theory in smart cities, across elements characterizing visibilities and invisibilities (InVisibilities) what emerges is ambience, change and appreciative inquiry, and AmI and complexity. For Chapter 6, methods and approaches, context and imagination appear as well as the importance of people and the presence of emotion, attuned, and ambiance. For Chapter 7, ambient approaches and inquiries, what emerges is the augmenting of well-being and livability awareness with many forms of data such as that for air quality along with implications for access to public data in real time, health data for personal monitoring on smartphones and watches, and so on.

Visually, observations from Table 1 are depicted in Fig. 2 as an overview of elements emerging from the literature review for visibilities and invisibilities

Figure 2. Elements emerging from the literature for visibilities and invisibilities in smart cities

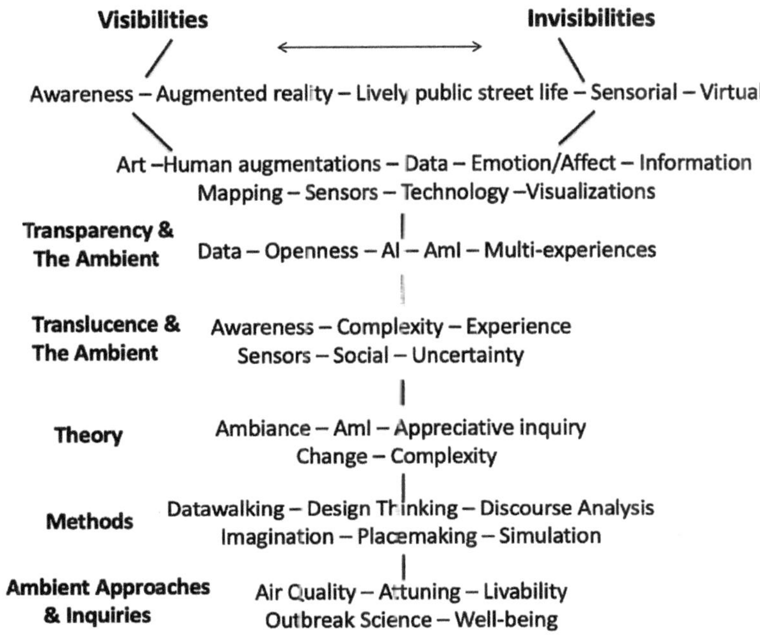

in smart cities. These terms give rise to, inform, and guide the selection of elements or proxies that were used in exploring relationships in smart cities in this work. In Chapter 1, the interweaving of visibilities and invisibilities as InVisibilities surfaces the elements of awareness, augmented reality, lively public street life, sensorial, and the virtual. In Chapter 2, making the invisible visible surfaces the elements of art, human augmentations, data, emotion/affect, information mapping, sensors, technology, and visualizations. In Chapter 3, transparencies and the ambient surfaces the elements of data, openness, artificial intelligence (AI), ambient intelligence (AmI), and multi-experiences. In Chapter 4, translucence and the ambient surfaces awareness, complexity, experience, sensors, social, and uncertainty.

In Chapter 5, exploring theory for visibilities and invisibilities, ambiance, AmI, appreciative inquiry, change, and complexity emerged. In Chapter 6, exploring methods and approaches for visibilities and invisibilities, datawalking, design thinking, discourse analysis, imagination, placemaking, and simulation emerged. In Chapter 7, exploring ambient approaches and inquiries, elements that emerged included air quality, attuning, livability, outbreak science, and well-being.

2.2 Research Questions and Propositions

Table 2 presents the research questions formulated for exploration in the seven chapters in the book.

Table 2. Research questions for visibilities and invisibilities in smart cities

Chapter	Research Question
1: Perspectives on Smart Cities	What is the nature of the relationship between visibilities and invisibilities in smart cities and why does this relationship matter?
2: Making the Invisible Visible	What does making the invisible visible mean in smart cities and why does this activity matter?
3: Transparency and the Ambient	What is the nature of the relationship between transparency and the ambient in smart cities and regions, as dimensions of visibilities and invisibilities?
4: Translucence and the Ambient	What is the nature of the relationship between translucence and the ambient in smart cities and regions, as dimensions of visibilities and invisibilities?
5: Theory in Smart Cities	Why does ambient theory matter for cities, the urban, and smart cities?
6: Methods in Smart Cities	Why does ambient methodology matter for cities, the urban, and smart cities?
7: Ambient Approaches & Inquiries	What is the nature of the relationship between the ambient & InVisibilities as well-being in smart cities & why does the ambient & well-being matter?

Table 3 presents the propositions, formulated from the research questions, for exploration in each of the seven chapters in the book.

2.3 Overview of Relationships

Table 4 presents the items correlated, and the Spearman correlation results for the ordinal data corresponding to the focus and dimensions for each of the seven chapters in the book.

In Chapter 1, exploring perspectives on smart cities, attuning to urban spaces as a proxy for the ambient was correlated with invisible infrastructures yielding a "very good" (Creswell, 2018) Spearman correlation of .82. Information and Communication Technologies (ICTs) were used as a proxy for invisible infrastructures and when correlated, a relationship "useful for limited prediction" (Creswell, 2018) was found as indicated by the Spearman correlation coefficient of .57.

In Chapter 2, exploring approaches to making the invisible visible, heightening urban sensibilities was correlated with urbanizing (e.g., adapting for urban uses) resulting in a relationship "useful for limited prediction"

Table 3. Propositions for visibilities and invisibilities in smart cities

Chapter	Propositions
1: Perspectives on Smart Cities	*Understanding the nature of the relationship between visibilities and invisibilities is important in smart cities and regions in relation to emergent and evolving spaces for aware people interacting with each other and with aware technologies*
2: Making the Invisible Visible	*Making the invisible visible in smart cities and regions provides greater understandings of the potentials enabled by heightening urban sensibilities for urbanizing, as in, adapting for urban uses, and emotion/affect such as feelings of tension or anger in gaining greater awareness capabilities for collaborating and interacting with each other and with aware technologies (digital/virtual) as opportunities important in evolving research and practice*
3: Transparency and the Ambient	*Insight into the relationship between transparency and the ambient in smart cities sheds light on the many dimensions of visibilities and invisibilities through the use of openness and attuning to urban spaces along with associated complexities and opportunities for research and practice*
4: Translucence and the Ambient	*Understanding the relationship between transparency and the ambient in smart cities enriches the dimensionalities of urban visibilities and invisibilities while highlighting associated complexities and opportunities*
5: Theory in Smart Cities	*Ambient theory matters for cities, the urban, and smart cities in accommodating explorations and implementations of emergent and evolving spaces for an interweaving of visibilities and invisibilities as InVisibilities in relation to the ambient*
6: Methods in Smart Cities	*Ambient methodologies matter for cities, urban regions, and smart cities in accommodating explorations and implementations of emergent and evolving spaces for an interweaving of visible and invisible dimensions as InVisibilities in relation to the ambient*
7: Ambient Approaches & Inquiries	*Understanding the nature of the relationship(s) between the ambient and InVisibilities as well-being is important in smart cities and regions in relation to emergent and evolving spaces for humanity and more specifically for aware people interacting with each other and with aware technologies*

(Creswell, 2018) with a Spearman correlation of .43. Emotion/affect using the proxy of anger was also explored in relation to awareness as a proxy for the ambient yielding a "very good" (Creswell, 2018) correlation of .81.

In Chapter 3, exploring the dimensions of transparency and the ambient, openness as a proxy for transparency was correlated with attuning to urban spaces resulting in a relationship "useful for limited prediction" (Creswell, 2018) with a Spearman correlation of .43.

In Chapter 4, exploring the dimensions of translucence and the ambient, the notion of visible infrastructures as a proxy for translucence was correlated with awareness resulting in a "slight relationship" (Creswell, 2018) with a Spearman correlation of .27.

In Chapter 5, exploring theory and the ambient, ICTs as a proxy for invisible infrastructures was correlated with awareness as a proxy for the

Table 4. Overview of correlations for visibilities and invisibilities in smart cities

Relationships for Variables Associated with Visibilities and Invisibilities in Smart Cities	
Chapter Focus and Variables Explored	**Correlation (Spearman)**
Chapter 1: Perspectives on Smart Cities *Attuning to Urban Spaces (the Ambient)* and *Invisible Infrastructures* *Visible Infrastructures* and *ICTs as Invisible Infrastructures*	.82 .57
Chapter 2: Making the Invisible Visible *Heightening urban sensibilities* and *Urbanizing (Adapting for Urban Uses)* *Emotion/affect (Anger)* and *Awareness (the Ambient)*	.43 .81
Chapter 3: Transparency and The Ambient *Openness* and *Attuning to Urban Spaces (the Ambient)*	.43
Chapter 4: Translucence and The Ambient *Visible Infrastructures* and *Awareness (the Ambient)*	.27
Chapter 5: Theory in Smart Cities *ICTs* and *Awareness (the Ambient)*	-.23
Chapter 6: Methods and Approaches *Placemaking* and *Attuning to Urban Spaces (the Ambient)*	.43
Chapter 7: Ambient Approaches and Inquiries *Air Quality (Livability)* and *Attuning to Urban Spaces (the Ambient)*	.80

ambient resulting in an inverse relationship (Creswell, 2018) with a Spearman correlation of -.23.

In Chapter 6, exploring methods and approaches and the ambient, placemaking as a proxy for invisible infrastructures was correlated with attuning to urban spaces as a proxy for the ambient resulting in a relationship "useful for limited prediction" (Creswell, 2018) with a Spearman correlation of .43.

In Chapter 7, exploring ambient approaches and inquiries, air quality as a proxy for livability was correlated with attuning to urban spaces as a proxy for the ambient yielding a "very good" (Creswell), 2018) correlation of .80.

In Chapters 1, 2, and 7 a "very good" correlation in the .66 to .85 range was achieved in this work, yet in other chapters, correlations occurred in the .35 to .65 range, said to be "useful for limited prediction" (Creswell, 2018) and important "to identify variable membership in the statistical procedure of factor analysis" as in "the intercorrelation of variables" pointing to potential directions for research and practice opportunities going forward. Chapter 4 shows "only a slight relationship" (Creswell, 2018) between translucence and the ambient at .27, a relationship that may evolve and develop further over time as these concepts gain greater attention and understanding. Chapter 5 shows a slight relationship for ICTs and awareness that is inverse at -.23 when explored in the context of theory in smart cities and this may evolve and change over time as the theoretical foundations for smart cities develops and strengthens over time.

2.4 Observations and Interpretations From Study Explorations

Table 5 brings together observations that emerged from study explorations for visibilities and invisibilities in smart cities corresponding to the focus and dimensions for each of the severn chapters in the book. For Chapter 1, perspectives on smart cities, key observations and interpretations pertain to tangible real-world examples of the interweaving of the physical and digital shedding light on why visibilities and invisibilities matter in smart cities. For Chapter 2, making the invisible visible, key observations and interpretations pertain to the need for more interactive urban displays with meaningful content focused on informing and learning; and that smart cities and regions involve much broader understandings of "smartness" extending to the human

Table 5. Observations from study explorations for visibilities and invisibilities in smart cities

Focus and Dimensions	Observations and Interpretations
1: Perspectives on Smart Cities *Visible Infrastructures* *Invisible Infrastructures (ICTs)*	• Tangible real-world examples of the interweaving of the physical & digital shedding light on why visibilities & invisibilities matter in smart cities
2: Making the Invisible Visible *Heightening Urban Sensibilities* *Urbanizing*	• The need for more interactive urban displays with meaningful content focused on informing & learning • Smart cities involve broader understandings of "smartness" extending to the human elements of heightening urban sensibilities, urbanizing, etc.
3: Transparency & The Ambient *Openness* *Attuning to Urban Spaces*	• Support for the notion by other researchers of use through participation & interaction and an open mindset
4: Translucence & The Ambient *Visible Infrastructures* *Awareness*	• Constraints, accountability & the social find expression as people voice their understandings of smart cities & their experiences become critical
5: Theory in Smart Cities *ICTs (InVisibilities)* *Awareness (the Ambient)*	• The inverse, slight relationship found between *ICTs & the ambient* may be pointing to the complexities & uncertainties of smart cities
6: Methods & Approaches *Place-making* *Attuning to Urban Spaces*	• Placemaking & design thinking reveal the emergence of platforms as physical & digital for viewing & sharing information, art, history & culture, contributing to learning • Data visualization seems to enable spaces to accommodate "smart cities" as "wise cities" considerating the need for fun, meaningfulness, learning
7: Ambient Approaches & Inquiries	• Human infrastructure: "multi layers of senses"; Digital: air quality data

elements of heightening urban sensibilities, urbanizing, and the identification and assessment of other potential variables.

For Chapter 3, transparency and the ambient, key observations and interpretations pertain to support for the notion by researchers (Matheus and Janssen, 2019) of use through participation and interaction and an open mindset (Mulder, 2015). For Chapter 4, translucence and the ambient, key observations and interpretations pertain to constraints, accountability, and the social that find expression as people voice their understandings of smart cities and it is thus that people and their experiences become critical in smart cities. For Chapter 5, on theory, key observations and interpretations pertain to the musing that it may be that the inverse, slight relationship found between *ICTs* (as a proxy for visibilities and invisibilities as *InVisibilities)* and *awareness* (as proxy for *the ambient)* highlights the complexities, uncertainties, and challenges associated with understanding of smart cities. For Chapter 6, on methods, key observations and interpretations pertain to approaches such as placemaking and design thinking that reveal the emergence of platforms as both physical and digital for the viewing and sharing of information, art, history and culture, contributing to learning and data visualization that would seem to enable spaces to accommodate "smart cities" as "wise cities" taking into consideration the needs of people for fun, meaningfulness, and learning. For Chapter 7, ambient approaches and inquiries, key observations and interpretations pertain to well-being and livability contributing to an interweaving of human and digital infrastructures.

2.5 Synthesis of Insights, Ideas, and Exercises

Table 6 provides a glimpse of key insights that emerged for visibilities and invisibilities in smart cities, corresponding to the focus and dimensions for each of the seven chapters in the book.

For Chapter 1, perspective on smart cities, key insights pertaining to visible and invisible infrastructures include: the presence of a promising relationship between ICTs as a proxy for invisible infrastructures when correlated with visibilities; the interweaving of the visible and invisible in smart urban environments/regions; and the unforeseen value in relation to the serendipitous generation, collection, and use of urban public data as a type of invisibility with latent visibility potential.

For Chapter 2, making the invisible visible, key insights pertain to the emerging range of approaches to making the invisible visible in evolving

Table 6. Synthesis of insights for visibilities and invisibilities in smart cities

Insights for Visibilities and Invisibilities in Smart Cities	
Focus and Dimensions	**Insights**
1: Perspectives on Smart Cities *Visible Infrastructures* *Invisible Infrastructures (ICTs)*	• Promising relationship between invisibilities (ICTs) and visibilities • Interweaving of the visible & invisible in smart urban environments • Unforeseen value for serendipitous generation, collection & use of urban public data as a type of invisibility with latent visibility potential
2: Making the Invisible Visible *Heightening Urban Sensibilities* *Urbanizing*	• Range of approaches to making the invisible visible in evolving understandings of smart cities • A "useful" correlation for *heightening urban sensibilities & urbanizing* • Interplay of augmentations, interactive displays & urban elements
3: Transparency & The Ambient *Openness* *Attuning to Urban Spaces*	• Dimensionality for invisibilities & visibilities as *transparency/ambient* • A "slight" relationship when *openness & attuning* were correlated • Interplay of data, design & experiences for *transparency & the ambient*
4: Translucence & The Ambient *Visible Infrastructures* *Awareness (the Ambient)*	• Dimensionality of invisibilities & visibilities as *translucence/ambient* • A "slight" relationship between *visible infrastructures* and *awareness* • Interplay of accountability, constraints, social for *translucence/ambient*
5: Theory in Smart Cities *ICTs (InVisibilities)* *Awareness (the Ambient)*	• Dimensionality for invisibilities & visibilities as *ICTs & awareness* • A "slight" inverse relationship for *ICTs* correlated with *awareness* Interplay of ambience, atmospheres, social for *InVisibilities & ambient*
6: Methods & Approaches *Placemaking* *Attuning to Urban Spaces*	• Dimensionality for invisibilities & visibilities as *placemaking & attuning* • "Limited prediction" for *placemaking & attuning to urban spaces* • Interplay of datawalking, data visualizations & design thinking
7: Ambient Approaches & Inquiries	• Potential for opportunities to assess *livability* in terms of *air quality* • Importance of assessments of *sharing non-sensitive data* • Surfacing of variables associated with *access to public data* • Presence of a range of relationships between *data & the ambient*

understandings of smart cities; the presence of a "useful" correlation between *heightening urban sensibilities* and *urbanizing*; and the interplay of augmentations, urban interactive displays, and urban elements in making visible the invisible.

For Chapter 3, transparency and the ambient, key insights pertain to the emerging range of dimensions for invisibilities and visibilities as *transparency* and *the ambient*; the presence of a "slight" relationship emerged when *openness* and *attuning to urban spaces* were correlated; and the interplay of data, design, and experiences associated with *transparency* and *the ambient.*

For Chapter 4, translucence and the ambient, key insights pertain to the emerging range of dimensions for invisibilities and visibilities as *translucence* and *the ambient*; the presence of a "slight" relationship emerged when *translucence* (e.g., *visible infrastructures)* was correlated with *the ambient* (e.g., *awareness*); and the interplay of accountability, constraints, and the social, associated with *translucence* and *the ambient.*

For Chapter 5, on theory, key insights pertain to the emerging range of dimensions for invisibilities and visibilities as *ICTs* (e.g., *InVisibilities)* and *the ambient* (e.g., *awareness*); the presence of a "slight" inverse relationship emerged when *InVisibilities* (e.g., *ICTs)* and *the ambient* (e.g., *awareness)* were correlated; and the interplay of ambience, atmospheres, and the social associated with *InVisibilities* and *the ambient.*

For Chapter 6, on methods and approaches, key insights pertain to the emerging range of dimensions for invisibilities and visibilities as *place-making* (e.g., *InVisibilities)* and *attuning to urban spaces* (e.g., *the ambient);* the presence of "limited prediction" emerged when *InVisibilities* (e.g., *place-making)* and *the ambient* (e.g., *attuning to urban spaces)* were correlated; and the interplay of datawalking, data visualizations, and design thinking as approaches to *InVisibilities* and *the ambient.*

For Chapter 7, ambient approaches and inquiries, key insights pertain to the potential for opportunities to assess *livability* in terms of *air quality*; the importance of assessments of *sharing non-sensitive data;* the surfacing of variables associated with *access to public data*; and the presence of a range of relationships between *data* and *the ambient.*

Table 7 provides an overview of key ideas that emerged for visibilities and invisibilities in smart cities corresponding to the focus and dimensions for each of the seven chapters in the book.

For Chapter 1, perspectives on smart cities, key ideas pertaining to visible and invisible infrastructures include: opportunities potential that invisible infrastructures as ICTs (information and communication technologies) present in smart cities; potential for people to engage more meaningfully with invisible spaces in digital/physical urban spaces; and opportunities potential for visibly and invisibly interacting with infrastructures.

For Chapter 2, making the invisible visible, key ideas pertain to the importance of people becoming more aware of the invisible and intangible aspects of the smart cities phenomenon; potential for people to shed light on rendering visible, invisible spaces in digital/physical urban environments; and the potential for opportunities going forward to involve people more meaningfully in making the invisible visible.

For Chapter 3, transparency and the ambient, key ideas pertain to the importance of people becoming more involved in understandings of *openness* and *attuning to urban spaces* in relation to *transparency* and *the ambient* as dimensions of invisibilities and visibilities in smart cities; the potential for people to shed light on *transparency* and *the ambient* in urban spaces; and explorations of *transparency* and *the ambient* as novel practice and research spaces for smarter cities and regions.

For Chapter 4, translucence and the ambient, key ideas pertain to the importance of people becoming more involved in understandings of *translucence* and *the ambient* as dimensions of invisibilities and visibilities; the potential for people to shed light on *translucence* and *the ambient* in urban

Table 7. Synthesis of ideas for visibilities and invisibilities in smart cities

Focus and Dimensions	Ideas
1: Perspectives on Smart Cities *Visible Infrastructures* *Invisible Infrastructures (ICTs)*	• Opportunities potential that invisible infrastructures as ICTs present in smart cities • Potential for people to engage more meaningfully with invisible spaces in digital/physical urban spaces • Opportunities potential for visibly & invisibly interacting with infrastructures
2: Making the Invisible Visible *Heightening Urban Sensibilities* *Urbanizing*	• Importance of people becoming more aware of the invisible & intangible aspects of the smart cities phenomenon • Potential for people to shed light on rendering visible, invisible spaces in digital/physical urban environments • Potential for opportunities going forward to involve people more meaningfully in making the invisible visible
3: Transparency & The Ambient *Openness* *Attuning to Urban Spaces*	• Importance of people becoming more involved in understandings of *openness & attuning* in relation to *transparency & the ambient* as dimensions of invisibilities & visibilities in smart cities • Potential for people to shed light on *transparency & the ambient* in urban spaces • Explorations of *transparency & the ambient* as novel practice & research spaces for smarter cities & regions
4: Translucence & The Ambient *Visible Infrastructures* *Awareness*	• Importance of people becoming more involved in understandings of *translucence & the ambient* as dimensions of invisibilities & visibilities • Potential for people to shed light on *translucence & the ambient* in urban spaces • Explorations of *translucence & the ambient* as novel practice & research spaces for smarter cities & regions
5: Theory in Smart Cities *ICTs* *Awareness*	• Importance of people becoming more involved in understandings of *ICTs & the ambient* as dimensions of invisibilities & visibilities • Potential for people to shed light on *ICTs & awareness* in urban spaces • Explorations of *ICTs & the ambient* as novel research spaces informing theory for smarter cities & regions
6: Methods & Approaches *Place-making* *Attuning to Urban Spaces*	• Importance of people becoming more involved in understandings of *place-making & the ambient* as dimensions of invisibilities & visibilities • Potential for people to shed light on *place-making & attuning to urban spaces* • Explorations of *place-making & the ambient* as novel research & practice spaces informing approaches & methodologies for smarter cities & regions
7: Ambient Approaches & Inquiries	• Assessments of livability as a way for people to become more involved • Potential for people to shed more light on the invisible in digital/physical urban spaces through assessments of their experiences with data • Possibly novel approaches to study & practice involving more aware & attuned people

spaces; and explorations of *translucence* and *the ambient* as novel practice and research spaces for smarter cities and regions.

For Chapter 5, on theory, key ideas pertain to the importance of people becoming more involved in understandings of *ICTs* and *the ambient* as dimensions of invisibilities and visibilities; the potential for people to shed light on *ICTs* and *awareness* in urban spaces; and explorations of *ICTs* and *the ambient* as novel research spaces informing theory for smarter cities and regions.

For Chapter 6, on methods and approaches, key ideas pertain to the importance of people becoming more involved in understandings of *placemaking* and *the ambient* as dimensions of invisibilities and visibilities; the potential for people to shed light on *place-making* and *attuning to urban spaces*; and explorations of *place-making* and *the ambient* as novel research and practice spaces informing approaches and methodologies for smarter cities.

For Chapter 7, ambient approaches and inquiries, key ideas pertain to the importance of assessments of livability as a way for people to become more involved; the potential for people to shed more light on invisible spaces in digital/physical urban spaces through assessments of their experiences of sharing non-sensitive data and accessing public data in their city; and ambient inquiries as possibly novel approaches to the study and practice of smarter cities involving more aware and attuned people.

Table 8 provides a series of exercises designed for the reader to consider, encouraging further explorations of visibilities and invisibilities in smart cities organized by focus and dimensions for each of the eight chapters in the book. As such, readers of this book are invited to further explore visibilities and invisibilities in their city through completing the exercises at the end of each chapter with an option to share their exercise responses with the author through an online form.

For Chapter 1, perspectives on smart cities, readers are asked to: identify one or more smart elements in their city that are visible to the public; identify one or more smart elements in their city that are invisible to the public; and describe one or more ways in which smart elements in their city that are invisible to the public could be meaningfully interacted with.

For Chapter 2, making the invisible visible, readers are asked to: identify one or more examples of making the invisible visible in their city; identify one or more approaches for making the invisible visible in their city based on their example; and describe why it could be important to make the invisible visible, based on their example.

For Chapter 3, transparency and the ambient, readers are asked to: identify one or more examples of openness and transparency in their city; identify one or more examples of attuning to urban spaces and the ambient in their

Table 8. Exercises encouraging further explorations of visibilities and invisibilities in smart cities

Focus and Dimensions	Exercises
1: Perspectives on Smart Cities *Visible Infrastructures* *Invisible Infrastructures (ICTs)*	1. Identify one or more smart elements in your city that are visible to the public 2. Identify one or more smart elements in your city that are invisible to the public 3. Describe one or more ways in which smart elements in your city that are invisible to the public could be meaningfully interacted with
2: Making the Invisible Visible *Heightening Urban Sensibilities* *Urbanizing*	1. Identify one or more examples of making the invisible visible in your city 2. Identify one or more approaches for making the invisible visible in your city based on your example 3. Describe why it could be important to make the invisible visible
3: Transparency & The Ambient *Openness* *Attuning to Urban Spaces*	1. Identify one or more examples of *openness & transparency* in your city 2. Identify one or more examples of *attuning to urban spaces & the ambient* in your city 3. Describe why *transparency & the ambient* could be important for smart cities, based on your example(s)
4: Translucence & The Ambient *Visible Infrastructures* *Awareness*	1. Identify one or more examples of *translucence in visible infrastructures* in your city 2. Identify one or more examples of *awareness & the ambient* in your city 3. Describe why *translucence & the ambient* could be important for smart cities, based on your example(s)
5: Theory in Smart Cities *ICTs* *Awareness*	1. Identify one or more examples of *InVisibilities & ICTs* in your city 2. Identify one or more examples of *the ambient & awareness* in your city 3. Describe why *ICTs & awareness* could be important for smart cities, based on your example(s)
6: Methods & Approaches *Place-making* *Attuning to Urban Spaces*	1. Identify one or more examples of *InVisibilities & placemaking* in your city 2. Identify one or more examples of *the ambient & attuning to urban spaces* in your city 3. Describe why *InVisibilities* as *place-making & attuning to urban spaces* could be important for smart cities, based on your example(s)
7: Ambient Approaches & Inquiries	1. Identify one or more examples of *well-being* and *attuning to urban spaces* in your city 2. Identify one or more examples of *livability* and *attuning to urban spaces* in your city 3. Describe why *sharing non-sensitive data* and *attuning to urban spaces* could be important for smart cities, based on your example(s) 4. Describe why *access to public data* and *attuning to urban spaces* could be important for smart cities, based on your example(s)
8: Synthesis & Taxonomy	1. Describe how the Taxonomy presented in this chapter could be helpful 2. Identify what may be missing from this Taxonomy 3. Identify what seems to be missing from this exploration of visibilities and invisibilities in smart cities in this book

city; and describe why transparency and the ambient could be important for smart cities, based on their example(s).

For Chapter 4, translucence and the ambient, readers are asked to: identify one or more examples of translucence in visible infrastructures and awareness in their city; identify one or more examples of awareness and the ambient in their city; and describe why translucence and the ambient could be important for smart cities, based on their example(s).

For Chapter 5, on theory, readers are asked to: identify one or more examples of inVisibilities and ICTs in their city; identify one or more examples of the ambient and awareness in their city; and describe why ICTs and awareness could be important for smart cities, based on their example(s).

For Chapter 6, on methods and approaches, readers are asked to: identify one or more examples of InVisibilities and place-making in their city; identify one or more examples of the ambient and attuning to urban spaces in their city; and describe why place-making and attuning to urban spaces could be important for smart cities, based on their example(s).

For Chapter 7, ambient approaches and inquiries, readers are asked to: identify one or more examples of *well-being* and *attuning to urban spaces* in their city; identify one or more examples of *livability* and *attuning to urban spaces* in their city; describe why *sharing non-sensitive data* and *attuning to urban spaces* could be important for smart cities, based on their example(s); describe why *access to public data* and *attuning to urban spaces* could be important for smart cities, based on their example(s)

For Chapter 8, synthesis and taxonomy, readers are asked to: describe how the Taxonomy presented in this chapter could be helpful; identify what may be missing from this Taxonomy; identify what seems to be missing from this exploration of visibilities and invisibilities in smart cities in this book.

3. GOING FORWARD

This section provides an overview of future directions in terms of opportunities for research and then for practice.

Table 9 provides the focus and dimensions for each of the seven chapters in the book in relation to the identification of future directions as opportunities for research that emerged for visibilities and invisibilities in smart cities.

For Chapter 1, visible and invisible infrastructures, opportunities for research emerge in relation to learning more about the capabilities and

Table 9. Overview of future directions as opportunities for research in smart cities

Focus and Dimensions	Opportunities
1: Perspectives on Smart Cities *Visible Infrastructures* *Invisible Infrastructures (ICTs)*	• Learning more about capabilities / potentials of interactive urban displays • Urban data visualization to sensing
2: Making the Invisible Visible *Heightening Urban Sensibilities* *Urbanizing*	• Augmentations extending understandings of urban visualizations • Urban interactive displays for awareness, learning & informing in real time
3: Transparency & The Ambient *Openness* *Attuning to Urban Spaces*	• Data initiatives focusing on improving the quality of the urban • Design attuned to "people & their needs" accommodating wise & smart
4: Translucence & The Ambient *Visible Infrastructures* *Awareness*	• Explorations of accountability in urban relationships • Explorations of constraints associated with cities & the social element
5: Theory in Smart Cities *ICTs* *Awareness*	• *Ambiences* awareness in public spaces - dynamic & interactive visualizations • *Atmospheres* fostered through attentiveness to the pulse, vibrancy & sense • *Complexities* extending meetings with tech options for interactions
6: Methods & Approaches *Place-making* *Attuning to Urban Spaces*	• *Education* fostered through collaborative urban initiatives • *Learning* about smart cities through meaningful real world data collection • *Engagement* fun & meaningful urban activities & data visualizations
7: Ambient Approaches & Inquiries	• Ambient Methodologies • Ambient InVisibilities and Well-being

potentials of interactive urban displays as visible infrastructures and about invisible infrastructures, ranging from urban data visualizations to sensing.

For Chapter 2, making the invisible visible, opportunities for research emerge in relation to augmentations extending understandings of urban visualizations through heightening urban sensibilities and urban interactive displays for awareness, learning, and informing in real time as urbanizing (e.g., adapting for urban uses).

For Chapter 3, understanding and leveraging transparency and the ambient, opportunities for research emerge in relation to data initiatives focusing on improving the quality of the urban with openness and the exploration of design attuned to "people and their needs" accommodating wisdom and smartness while attuning to urban spaces.

For Chapter 4, understanding and leveraging translucence and the ambient, opportunities for research emerge in relation to explorations of accountability in urban relationships as visible infrastructures and the undertaking of explorations of constraints associated with any given city and the importance of the social element in support of awareness. For Chapter 5, on theory, opportunities for research emerge in relation to ICTs as *ambiences* through urban initiatives focusing on awareness in public spaces with dynamic,

Table 10. Overview of future directions as opportunities for practice in smart cities

Focus and Dimensions	Opportunities
1: Perspectives on Smart Cities *Visible Infrastructures* *Invisible Infrastructures (ICTs)*	• Technology infrastructures fostering interactive public spaces – data visualizations • Human infrastructures fostering more dynamic interactivities – social media, etc.
2: Making the Invisible Visible *Heightening Urban Sensibilities* *Urbanizing*	• Use of urban interactive displays, urban elements & augmentations • Use of heightening urban sensibilities & urbanizing as smart city indicators
3: Transparency & The Ambient *Openness* *Attuning to Urban Spaces*	• Use of data, design & experiences in relation to openness • Exploration of data, design & privacy in relation to attuning to urban spaces
4: Translucence & The Ambient *Visible Infrastructures* *Awareness*	• Use of urban displays, experiences & aware technologies • Urban initiatives: more dynamic, adaptive & responsive for problem-solving
5: Theory in Smart Cities *ICTs* *Awareness*	• Cross-sector participation in smart city initiatives • More dynamic, adaptive & responsive urban problem-solving
6: Methods & Approaches *Place-making* *Attuning to Urban Spaces*	• Potentials associated with place-making, datawalking & data visualizations • Broad cross-sector participation in the development, design & implementation of urban initiatives
7: Ambient Approaches & Inquiries	• Learning more about variables for livability & well-being in smart cities

and interactive visualizations; *atmospheres* fostered through attentiveness to the pulse, vibrancy, and sense of the city; and *complexities* understood through extending meetings with technology-based options for interactions. For Chapter 6, on methods, opportunities for research emerge in relation to place-making and attuning to urban spaces as *education* fostered through urban initiatives involving educational institutions, communities, and local government; *learning* about smart cities through meaningful real world data collection and analysis activities using mobile apps and other platforms; and *engagement* fostered through fun and meaningful urban activities and data visualizations. For Chapter 7, ambient approaches and inquiries, opportunities for research emerge in relation to ambient methodologies and the conceptual framework for augmented awareness in terms of further testing and validation for well-being and other variables.

Table 10 provides the focus and dimensions for each of the seven chapters in the book in relation to the identification of future directions as opportunities for practice that emerged for visibilities and invisibilities in smart cities. For Chapter 1, visible and invisible infrastructures, opportunities for practice emerge in relation to technology infrastructures fostering interactive public spaces utilizing data visualization and human infrastructures fostering more dynamic interactivities through social media and other awareness-enabling spaces.

For Chapter 2, making the invisible visible, opportunities for practice emerge in relation to the use of urban interactive displays and other urban elements and augmentations while employing heightening urban sensibilities and urbanizing (e.g., adapting for urban uses) as smart city and region indicators. For Chapter 3, understanding and leveraging transparency and the ambient, opportunities for practice emerge in relation to the use of data, design, and experience associated with openness and the exploration of data, design, and privacy in relation to attuning to urban spaces.

For Chapter 4, understanding and leveraging translucence and the ambient, opportunities for practice emerge in relation to the use of urban displays, experiences, and aware technologies as visible infrastructures and the undertaking of urban initiatives that are more dynamic, adaptive, and responsive for problem-solving in support of awareness. For Chapter 5, on theory, opportunities for practice emerge in relation to cross-sector participation in smart city initiatives through ICTs and other aware technologies and the undertaking of more dynamic, adaptive, and responsive urban problem-solving in support of awareness. For Chapter 6, on methods, opportunities for practice emerge in relation to potential associated with place-making, datawalking, and data visualizations and the undertaking of broad cross-sector participation in the development, design, and implementation of urban initiatives attentive and attuned to urban spaces. For Chapter 7, ambient approaches and inquiries, opportunities for practice emerge in relation to learning more about variables for well-being and livability in smart environments.

4. TAXONOMY OF VISIBILITIES AND INVISIBILITIES IN SMART CITIES

Seels (1997) argues for taxonomic development, claiming that "in order for conceptual structures to evolve to theoretical systems, better taxonomic classification is a necessary first step." As such, where earlier chapters provided explorations of perspectives, dimensions, and aspects characterizing visibilities and invisibilities, the synthesis in this chapter enables formulation of a taxonomy of visibilities and invisibilities in smart cities as depicted in Fig. 3 in support of theory for smart cities.

Figure 3. Taxonomy of visibilities and invisibilities in support of theory for smart cities

Taxonomy

Invisibilities – Visibilities

Perspectives	Approaches	Domains	Variables
Creativity/Innovation	Awareness	Architecture	Access to Public Data
Data Science	Design Thinking	Information Technology	Air Quality
Engineering	Emotion/Affect	Law	ICTs
Health	Human-in-the-loop	Network Science	Livability
Human Geography	Participatory Design	Public Administration	Sharing of Non-personal Data
Information Behavior	Sensibilities	Sociology	Social Media
Learning	Urbanizing	Urban Planning	Well-being

Methods	Theory	The Ambient	Translucence	Transparency
Imagination	AmI	AI	Accessibility	Algorithms
	Ambiance/Atmospheres	Attuning	Experiences	Data
	Appreciative Inquiry	Augmented	Materials	Experience
Geospatial Analysis	Change	Awareness	Placemaking	Governance
Mapping	Complexity	CI/CN	Social	Infrastructures
Simulation	Discourse	Digital/Sensors	Socio-technical	Openness
Visualizations	Window	Material	Usability	Privacy

The taxonomy highlights the interwoven nature of invisibilities and visibilities where perspectives from the research and practice literature range from creativity and innovation to information behavior and learning. Approaches to working with invisibilities and visibilities range from that of awareness and design thinking to sensibilities and urbanizing. Domains for invisibilities and visibilities are many and varied, ranging from architecture and information technology to sociology and urban planning. Variables for the study, practice, and exploration of invisibilities and visibilities range from access to public data and air quality to social media and well-being.

Explorations of invisibilities and visibilities reveal dimensions such as transparency, translucence, and the ambient. The dimension of transparency gives rise to challenges and opportunities associated with many elements from algorithms and data to openness and privacy. The dimension of translucence gives rise to challenges and opportunities associated with many elements from accessibility and experiences to the socio-technical and usability. The dimension of the ambient gives rise to challenges and opportunities associated with many elements from artificial intelligence (AI) and attuning to the material and computational intelligence and communication networks (CI/CN). All of which gives rise to the need for theory and methods for invisibilities and visibilities in smart environments. Theory considered to be relevant for the study and practice of invisibilities and visibilities ranges from ambient intelligence (AmI) and change to discourse and window theory. Methods considered to be relevant for the study and practice of invisibilities and visibilities range from imagination and geospatial analysis to simulation and visualizations. It is worth noting that ambiance and atmospheres as well as appreciative inquiry are considered to be theory as well as methods.

5. CONCLUSION

Where earlier chapters provided explorations of perspectives, dimensions, and aspects of visibilities and invisibilities, formulating conceptual frameworks, and identifying challenges and opportunities, this chapter assembles the findings into a coherent whole. The synthesis presented in this chapter enables further opportunity for reflecting on the patterns and relationships found throughout this work. As part of this synthesis, a conceptual framework for visibilities, invisibilities, and the ambient in smart cities and regions is advanced, providing further enriching, broadening, and extending of the framework for ambient in/visibilities in smart city urbanities (McKenna, 2019). This revised and extended framework supports the potential for evolving urban theory to encompass ambient theory in smart environments in supporting understandings of smart cities and their potential while serving to possibly enrich Berman's (1982) notion that "all that is solid melts into air." The synthesis provided in this chapter also enables formulation of a taxonomy of visibilities and invisibilities in smart cities contributing further to theoretical foundations.

Limitations of this book related to literature review coverage on the one hand and sample size on the other for the study underlying this work are mitigated by the potential to extend coverage for the literature review and extend study coverage to larger cities in more countries. Additionally, limitations associated with the use of proxies in conducting correlations for various dimensions of visibilities and invisibilities are mitigated by the potential to conduct correlations more directly, identify additional variables to use for correlations, and continue testing existing correlations as perspectives evolve for smart cities and as initiatives and implementations become smarter in terms of involving more people in meaningful collaborations.

Reflections

Ever-present during the writing of this book is the global COVID-19 pandemic as evidenced in the various chapters that make reference to emerging concerns, either directly with reference to discussions of urban density and disease in Chapter 5, or indirectly through the inclusion of references in the Additional Readings section for each chapter. Indeed, Chapter 7 seeks to address more directly the opportunities enabled for practice and research through emerging reflections on uncertainties such as how the COVID-19 pandemic is said to be contributing to rapid transformations for living and working in cities; for designing cities going forward; and for evolving understanding of smart cities.

Critically important to any discussion of visibilities and invisibilities in smart cities are concerns with equality and inclusion, some of which are highlighted by the COVID-19 pandemic during the writing of this work. Other opportunities emerging from this work include the potential for further exploration of Anderson's (2011) body insight scale (BIS), self-relevant research (Amabile and Hall, 2019), together with the use of some of the variables identified in this book such as heightening human sensibilities, well-being, awareness, and attuning to urban spaces that may assist in opening spaces for action, practice, and research, involving people more meaningfully, going forward.

Exercises for the reader to consider, encouraging further exploration include the following:

1. Describe how the Taxonomy presented in this chapter could be helpful
2. Identify what may be missing from this Taxonomy
3. Identify what seems to be missing from this exploration of visibilities and invisibilities in smart cities in this book

If you would like to share your responses to the Exercises with the author, please complete the questions in this online space for Chapter 8 – A Synthesis and Taxonomy

This work will be of interest to a broad audience including educators, students, researchers, artists, urban planners, urban practitioners, community members, designers, developers, and anyone concerned with improving urban areas, in the face of complexities, uncertainties, and transformations contributing to unprecedented challenges and opportunities posed by working with the invisible and visible in support of smarter, more responsive cities and regions.

REFERENCES

Amabile, T. M., & Hall, D. T. (2019). The undervalued power of self-relevant research: The case of researching retirement while retiring. *Academy of Management Journal*. Advance online publication. doi:10.5465/amp.2018.0083

Anderson, R. (2011). Body insight scale. *Mind Garden Blog*. Retrieved 31 May 2020 from https://www.mindgarden.com/73-body-insight-scale

Berman, M. (1982). *All that is solid melts into air: The experience of modernity*. Simon & Schuster.

Creswell, J. W. (2018). *Educational Research: Planning, Conducting, and Evaluating Quantitative and Qualitative Research* (6th ed.). Pearson.

Matheus, R., & Janssen, M. (2019). A systematic literature study to unravel transparency enabled by open government data: The window theory. *Public Performance & Management Review*. Advance online publication. doi:10.1080/15309576.2019.1691025

McKenna, H. P. (2019). Visibilities and invisibilities: Methodologies and theoretical spaces for ambient urbanities. In *Ambient Urbanities as the Intersection between the IoT and the IoP in Smart Cities* (pp. 192–222). IGI Global. doi:10.4018/978-1-5225-7882-6.ch007

Mulder, I. (2015). Opening up: Towards a sociable smart city. In *Citizen's Right to the Digital City*. Springer. doi:10.1007/978-981-287-919-6_9

Seels, B. (1997). Taxonomic issues and the development of theory in instructional technology. *Educational Technology*, *37*(1), 12–21.

ADDITIONAL READING

Wiener, S., & Iton, A. (2020). A backlash against cities would be dangerous: Undue fears of urban density warp public policy – and make Americans more vulnerable. *The Atlantic*. Retrieved 17 May 2020 from https://www.theatlantic.com/ideas/archive/2020/05/urban-density-not-problem/611752/

KEY TERMS AND DEFINITIONS

Ambient InVisibilities: Ambient InVisibilities refers to the increasing presence of more aware people together with awareness-enabled technologies in support of interactivity in the context of environments characterized by an interweaving of visibilities and invisibilities.

Ambient Methodologies: Ambient methodologies refer to more adaptive approaches to explorations of awareness in relation to people and awareness-enabled and enabling technologies encompassing the physical, digital, and less tangible, but not limited to ambiences, atmospheres, augmentations, emotion/affect, and the like.

Ambient Theory: Ambient theory refers to an emerging new addition to existing theories for urban research that is more amenable to adaptive frameworks for smart, responsive, and future cities while enabling possibly more aware, dynamic, interactive, and evolving models in response to increasing complexities and uncertainties.

Awareness: Awareness is conceptualized in this work as the quality of being aware as it applies to people on the one hand, to the enabling of technologies on the other, and to the enhancing of awareness through people interacting with awareness-enabled and enabling technologies.

Invisibilities: Invisibilities refer to elements that are rendered invisible or that may present as intangibles.

Smart Cities: Smart cities are urban areas and regions characterized by more aware and engaged people, interacting with each other and aided by the use of more awareness enhancing technologies for a wide range of purposes from communication to design to atmospheres to mobility to livability, to governance, to data generation and use for learning and informed decision-making.

Visibilities: Visibilities refer to physical and other visible urban elements and infrastructures.

Related Readings

To continue IGI Global's long-standing tradition of advancing innovation through emerging research, please find below a compiled list of recommended IGI Global book chapters and journal articles in the areas of smart cities, human geography, and sustainable urban development. These related readings will provide additional information and guidance to further enrich your knowledge and assist you with your own research.

Abalo, J. (2018). Is Street Begging a Result of Poverty, Budding Trade, or the Manipulation of Sentiments by Some Vested Interest Groups? In J. Mugambwa & M. W. Katusiimeh (Eds.), *Handbook of Research on Urban Governance and Management in the Developing World* (pp. 172–191). IGI Global. doi:10.4018/978-1-5225-4165-3.ch010

Abastante, F., Lombardi, P., & Moghadam, S. T. (2019). Defining Energy Criteria in the Absence of Open Data. In A. Voghera & L. La Riccia (Eds.), *Spatial Planning in the Big Data Revolution* (pp. 139–160). IGI Global. doi:10.4018/978-1-5225-7927-4.ch006

Abburu, S. (2019). Geospatial Semantic Query Engine for Urban Spatial Data Infrastructure. *International Journal on Semantic Web and Information Systems, 15*(4), 31–51. doi:10.4018/IJSWIS.2019100103

Abdullahi, R. B. (2018). Exploring the Crowdfunding of Nigerian Volunteer Corps' Efforts in Urban Development. In U. G. Benna & A. U. Benna (Eds.), *Crowdfunding and Sustainable Urban Development in Emerging Economies* (pp. 277–298). IGI Global. doi:10.4018/978-1-5225-3952-0.ch014

Abdullahi, R. B. (2018). Volunteerism in Urban Development the Case of Non-Cash, Non-Digital Crowdfunding Growth in Nigeria. In U. G. Benna & A. U. Benna (Eds.), *Crowdfunding and Sustainable Urban Development in Emerging Economies* (pp. 188–210). IGI Global. doi:10.4018/978-1-5225-3952-0.ch010

Abedo, M., Salheen, M., & Elshater, A. (2020). The Dawn of Walkability. In R. M. Doheim, A. A. Farag, & E. Kamel (Eds.), *Humanizing Cities Through Car-Free City Development and Transformation* (pp. 98–114). IGI Global. doi:10.4018/978-1-7998-3507-3.ch004

Aboagye, D., Attakora-Amaniampong, E., & Owusu-Sekyere, E. (2020). Place-Based Assessment of Intersection of Biophysical and Social Vulnerability to Flooding in Accra, Ghana. *International Journal of Applied Geospatial Research, 11*(1), 55–68. doi:10.4018/IJAGR.2020010104

Abusaada, H. (2019). Revisiting the Word "Atmospheres" in the Urban Design Academic Field Based on the Artworks. *International Journal of Public and Private Perspectives on Healthcare, Culture, and the Environment, 3*(2), 22–42. doi:10.4018/IJPPPHCE.2019070102

Abusaada, H. (2020). Cinematic Works-Based Affective Urban Design Atmospheres. In M. Merviö (Ed.), *Global Issues and Innovative Solutions in Healthcare, Culture, and the Environment* (pp. 120–132). IGI Global. doi:10.4018/978-1-7998-3576-9.ch007

Abusaada, H. (2020). The Atmospheres of Urban Environments. In H. Abusaada, A. M. Salama, & A. Elshater (Eds.), *Reconstructing Urban Ambiance in Smart Public Places* (pp. 1–22). IGI Global. doi:10.4018/978-1-7998-3856-2.ch001

Abusaada, H., & Elshater, A. (2018). Knowledge-Based Urban Design in the Architectural Academic Field. In A. A. Alraouf (Ed.), *Knowledge-Based Urban Development in the Middle East* (pp. 204–227). IGI Global. doi:10.4018/978-1-5225-3734-2.ch011

Abusaada, H., & Elshater, A. (2019). A Crowdsourcing Approach in Urban Design. In H. Abusaada, C. Vellguth, & A. Elshater (Eds.), *Handbook of Research on Digital Research Methods and Architectural Tools in Urban Planning and Design* (pp. 1–24). IGI Global. doi:10.4018/978-1-5225-9238-9.ch001

Abusaada, H., & Elshater, A. (2019). From Chaos to Order. In U. G. Benna (Ed.), *Industrial and Urban Growth Policies at the Sub-National, National, and Global Levels* (pp. 41–64). IGI Global. doi:10.4018/978-1-5225-7625-9.ch003

Abusaada, H., & Elshater, A. (2019). Internet of Things Interpreting a Myth. In H. Abusaada, C. Vellguth, & A. Elshater (Eds.), *Handbook of Research on Digital Research Methods and Architectural Tools in Urban Planning and Design* (pp. 47–81). IGI Global. doi:10.4018/978-1-5225-9238-9.ch004

Abusaada, H., & Elshater, A. (2020). Affective Atmospheres, Essence of Architecture, and Spirit of Place. In H. Abusaada, A. M. Salama, & A. Elshater (Eds.), *Reconstructing Urban Ambiance in Smart Public Places* (pp. 41–59). IGI Global. doi:10.4018/978-1-7998-3856-2.ch003

Abusaada, H., & Elshater, A. (2020). Situational Atmospheres. In H. Abusaada, A. M. Salama, & A. Elshater (Eds.), *Reconstructing Urban Ambiance in Smart Public Places* (pp. 221–245). IGI Global. doi:10.4018/978-1-7998-3856-2.ch011

Abusaada, H., & Elshater, A. (2021). Addressing the New Pragmatic Methods in Urban Design Discipline. In Khosrow-Pour D.B.A., M. (Ed.), Encyclopedia of Information Science and Technology, Fifth Edition (pp. 1196-1213). IGI Global. doi:10.4018/978-1-7998-3479-3.ch083

Adad, M. A., Semlali, E. H., & Ibannain, F. (2019). Comprehensive Ontology Model of Moroccan Land Administration Domain Applied to the National Land Governance. In M. El-Ayachi & L. El Mansouri (Eds.), *Geospatial Technologies for Effective Land Governance* (pp. 28–47). IGI Global. doi:10.4018/978-1-5225-5939-9.ch003

Adesola, A. P., & Olla, G. O. (2019). Bridging the Digital Divide in Nigerian Information Landscape. *International Journal of Digital Literacy and Digital Competence*, *10*(3), 10–31. doi:10.4018/IJDLDC.2019070102

Adjei, P., & Montasari, R. (2020). A Critical Overview of Digital Twins. *International Journal of Strategic Engineering*, *3*(1), 51–61. doi:10.4018/IJoSE.2020010104

Adkisson, A. C., & Monaghan, C. H. (2018). Critical and Transformational Perspectives on Career and Technical Education in the Twenty-First Century for Urban Adult Learners. In V. X. Wang (Ed.), *Critical Theory and Transformative Learning* (pp. 57–71). IGI Global. doi:10.4018/978-1-5225-6086-9.ch005

Ahmad, N. (2020). How to Eliminate Slums in Pakistan. *International Journal of Urban Planning and Smart Cities, 1*(2), 30–42. doi:10.4018/IJUPSC.2020070103

Ahmed, N. O. (2020). Contemplating the Street as a Place. In R. M. Doheim, A. A. Farag, & E. Kamel (Eds.), *Humanizing Cities Through Car-Free City Development and Transformation* (pp. 1–31). IGI Global. doi:10.4018/978-1-7998-3507-3.ch001

Ahuja, K., & Khosla, A. (2019). A Framework to Develop a Zero-Carbon Emission Sustainable Cognitive City. In K. Ahuja & A. Khosla (Eds.), *Driving the Development, Management, and Sustainability of Cognitive Cities* (pp. 1–26). IGI Global. doi:10.4018/978-1-5225-8085-0.ch001

Ahuja, S. P., & Wheeler, N. (2020). Architecture of Fog-Enabled and Cloud-Enhanced Internet of Things Applications. *International Journal of Cloud Applications and Computing, 10*(1), 1–10. doi:10.4018/IJCAC.2020010101

Al Shidhani, A. A. (2019). Cyber Defense Maturity Levels and Threat Models for Smart Cities. *International Journal of Information Security and Privacy, 13*(2), 32–46. doi:10.4018/IJISP.2019040103

Alashi, A. T., & Koramaz, T. K. (2019). Towards City Information Modeling. In H. Abusaada, C. Vellguth, & A. Elshater (Eds.), *Handbook of Research on Digital Research Methods and Architectural Tools in Urban Planning and Design* (pp. 105–120). IGI Global. doi:10.4018/978-1-5225-9238-9.ch006

Alberti, A. M. (2018). The NovaGenesis Smart Cities Model. In J. J. Rodrigues & A. Gawanmeh (Eds.), *Cyber-Physical Systems for Next-Generation Networks* (pp. 163–182). IGI Global. doi:10.4018/978-1-5225-5510-0.ch008

Alcaide-Muñoz, L., Alcaide-Muñoz, C., & Rodríguez Bolívar, M. P. (2020). Analyzing of the Evolution of the Field of E-Government and Trending Research Topics. In Digital Government and Achieving E-Public Participation (pp. 1-19). IGI Global. doi:10.4018/978-1-7998-1526-6.ch001

Alfaya, L., Muniz, P., Wilkes, D., Martinez, A., & Fernandez, C. (2020). Planning Mobility on Transboundary Shrinking Towns. *International Journal of E-Planning Research, 9*(4), 61–77. doi:10.4018/IJEPR.2020100104

Alfiero, S., Cane, M., Doronzo, R., & Esposito, A. (2018). Environmental Efficiency Analysis of Bus Transport in Italy. In L. C. Carvalho & E. Truant (Eds.), *Maintaining Sustainable Accounting Systems in Small Business* (pp. 24–45). IGI Global. doi:10.4018/978-1-5225-5267-3.ch002

Alfiky, M. A., & Raumer, H. S. (2019). GeoDesign and Urban Impact Analysis. In H. Abusaada, C. Vellguth, & A. Elshater (Eds.), *Handbook of Research on Digital Research Methods and Architectural Tools in Urban Planning and Design* (pp. 82–104). IGI Global. doi:10.4018/978-1-5225-9238-9.ch005

Alhakkak, N. M. (2019). BigGIS With Hadoop in MapReduce Environment. In H. Abusaada, C. Vellguth, & A. Elshater (Eds.), *Handbook of Research on Digital Research Methods and Architectural Tools in Urban Planning and Design* (pp. 25–32). IGI Global. doi:10.4018/978-1-5225-9238-9.ch002

Alhakkak, N. M. (2020). Towards a Smart Neighbourhoods Simulated Model. In H. Abusaada, A. M. Salama, & A. Elshater (Eds.), *Reconstructing Urban Ambiance in Smart Public Places* (pp. 114–125). IGI Global. doi:10.4018/978-1-7998-3856-2.ch006

Ali Arbab, I. T. (2021). Towards Creative Mixed-Use Axes in the Upper Middle Class Residential Zones in Cairo Metropolitan Context. In A. A. Galaby & A. A. Abdrabo (Eds.), *Handbook of Research on Creative Cities and Advanced Models for Knowledge-Based Urban Development* (pp. 217–239). IGI Global. doi:10.4018/978-1-7998-4948-3.ch011

Alizadeh, T., Farid, R., & Willems, L. (2018). The Role of Social Media in Public Involvement. In C. Nunes Silva (Ed.), *New Approaches, Methods, and Tools in Urban E-Planning* (pp. 310–342). IGI Global. doi:10.4018/978-1-5225-5999-3.ch011

AlNuaimi, S. F., Mohammed, W. M., & Ismael, N. T. (2019). The Governance of Transportation and Urban Design in Bahrain According to the Fourth Generation Industrial Revolution. *International Journal of Technology Diffusion*, *10*(3), 18–32. doi:10.4018/IJTD.2019070102

Alraouf, A. A. (2018). The Knowledge Urbanity in Qatar. In A. A. Alraouf (Ed.), *Knowledge-Based Urban Development in the Middle East* (pp. 60–79). IGI Global. doi:10.4018/978-1-5225-3734-2.ch004

Alraouf, A. A. (2018). The Myth of Knowledge-Based Urban Development in the Middle Eastern Cities. In A. A. Alraouf (Ed.), *Knowledge-Based Urban Development in the Middle East* (pp. 1–23). IGI Global. doi:10.4018/978-1-5225-3734-2.ch001

Althobaiti, O. S. (2020). IoT Limitations and Concerns Relative to 5G Architecture. In J. J. Thomas, U. Fiore, G. P. Lechuga, V. Kharchenko, & P. Vasant (Eds.), *Handbook of Research on Smart Technology Models for Business and Industry* (pp. 45–69). IGI Global. doi:10.4018/978-1-7998-3645-2.ch003

Amador, H. C. (2018). Contemporary Urbanizations in Public Water Reservoirs. In I. D. Rosa, J. C. Lopes, R. Ribeiro, & A. Mendes (Eds.), *Handbook of Research on Methods and Tools for Assessing Cultural Landscape Adaptation* (pp. 249–275). IGI Global. doi:10.4018/978-1-5225-4186-8.ch010

Ambrosini, L., Bassolino, E., & Scarpati, F. (2018). Thermal-Perception-Driven Adaptive Design for Wellbeing in Outdoor Public Spaces. In F. Aletta & J. Xiao (Eds.), *Handbook of Research on Perception-Driven Approaches to Urban Assessment and Design* (pp. 206–239). IGI Global. doi:10.4018/978-1-5225-3637-6.ch009

Ampatzidou, C., & Gugerell, K. (2019). Mapping Game Mechanics for Learning in a Serious Game for the Energy Transition. *International Journal of E-Planning Research, 8*(2), 1–23. doi:10.4018/IJEPR.2019040101

Angelucci, F., Di Girolamo, C., & Zazzero, E. (2018). New Designing Codes for Urban Infrastructures. In G. Carlone, N. Martinelli, & F. Rotondo (Eds.), *Designing Grid Cities for Optimized Urban Development and Planning* (pp. 209–237). IGI Global. doi:10.4018/978-1-5225-3613-0.ch012

Anunciação, P. F., & Peñalver, A. J. (2017). Information Urbanistic Perspective in the Context of Blue Economy. *International Journal of Sociotechnology and Knowledge Development, 9*(3), 65–83. doi:10.4018/IJSKD.2017070104

Apugo, D. L., Wendt, J. L., Deksissa, T., & Peters, M. (2018). The Science Behind Support. In Changing Urban Landscapes Through Public Higher Education (pp. 97-115). IGI Global. doi:10.4018/978-1-5225-3454-9.ch005

Aramendia-Muneta, M. E. (2020). European Capital of Smart Tourism. In L. C. Carvalho, L. Calisto, & N. Gustavo (Eds.), *Strategic Business Models to Support Demand, Supply, and Destination Management in the Tourism and Hospitality Industry* (pp. 255–265). IGI Global. doi:10.4018/978-1-5225-9936-4.ch014

Arieli, D. (2019). The Nazi Phantom. In N. Davidovitch, R. A. Cohen, & E. Lewin (Eds.), *Post-Holocaust Studies in a Modern Context* (pp. 85–111). IGI Global. doi:10.4018/978-1-5225-6258-0.ch005

Aririguzoh, S. A., Mogaji, E., & Odiboh, O. O. (2019). The Effect of Celebrity Endorsements on Consumers' Buying Behavior in South West Nigeria. In A. Gbadamosi (Ed.), *Exploring the Dynamics of Consumerism in Developing Nations* (pp. 176–194). IGI Global. doi:10.4018/978-1-5225-7906-9.ch008

Arisha, A. R., & El-Moneim, N. M. (2019). Space Syntax Beyond Cairo Street Markets. In H. Abusaada, C. Vellguth, & A. Elshater (Eds.), *Handbook of Research on Digital Research Methods and Architectural Tools in Urban Planning and Design* (pp. 250–281). IGI Global. doi:10.4018/978-1-5225-9238-9.ch012

Arnoldi, E., Bosua, R., Cooper, V., Greenfield, C., & Ch'ng, H. Y. (2021). Entrepreneurial Needs for Outer Urban Co-Working Hubs. In Y. Blount & M. Gloet (Eds.), *Anywhere Working and the Future of Work* (pp. 190–219). IGI Global. doi:10.4018/978-1-7998-4159-3.ch008

Arora, J. B., & Kaushik, S. (2020). IoT in Education. In R. Singh, A. Gehlot, V. Jain, & P. K. Malik (Eds.), *Handbook of Research on the Internet of Things Applications in Robotics and Automation* (pp. 300–317). IGI Global. doi:10.4018/978-1-5225-9574-8.ch015

Artopoulos, G., Charalambous, P., & Wehmeier, C. E. (2019). Immersive Computing and Crowd Simulation Techniques in Modelling Urban Commons. *International Journal of E-Planning Research*, *8*(1), 35–49. doi:10.4018/IJEPR.2019010103

Ates, M. (2020). Social Innovation as a New Social Policy Tool for Regional Government Institutions in Smart Territories. In J. M. Palma-Ruiz, J. Saiz-Álvarez, & Á. Herrero-Crespo (Eds.), *Handbook of Research on Smart Territories and Entrepreneurial Ecosystems for Social Innovation and Sustainable Growth* (pp. 53–66). IGI Global. doi:10.4018/978-1-7998-2097-0.ch004

Ates, M. (2020). The Role of Smart City Solutions on the Road to Smart Territories. In J. M. Palma-Ruiz, J. Saiz-Álvarez, & Á. Herrero-Crespo (Eds.), *Handbook of Research on Smart Territories and Entrepreneurial Ecosystems for Social Innovation and Sustainable Growth* (pp. 1–18). IGI Global. doi:10.4018/978-1-7998-2097-0.ch001

Attiwill, S. (2020). Urban Interiors and Interiorities. In L. Crespi (Ed.), *Cultural, Theoretical, and Innovative Approaches to Contemporary Interior Design* (pp. 58–67). IGI Global. doi:10.4018/978-1-7998-2823-5.ch002

Awan, J. H., Memon, S., Shah, A. A., & Pathan, K. T. (2020). Proposed Framework of Smart Transportation in Pakistan. *International Journal of Cyber Warfare & Terrorism*, *10*(4), 48–63. doi:10.4018/IJCWT.2020100104

Ayaz, N., & Akay, B. (2020). Smart Municipalities in Tourism. In E. Çeltek (Ed.), *Handbook of Research on Smart Technology Applications in the Tourism Industry* (pp. 391–413). IGI Global. doi:10.4018/978-1-7998-1989-9.ch018

Badawi, A. M. (2021). Urban Transformation and Creativity. In A. A. Galaby & A. A. Abdrabo (Eds.), *Handbook of Research on Creative Cities and Advanced Models for Knowledge-Based Urban Development* (pp. 102–119). IGI Global. doi:10.4018/978-1-7998-4948-3.ch006

Bagnato, V. P. (2018). Technology and Urban Structure. In G. Carlone, N. Martinelli, & F. Rotondo (Eds.), *Designing Grid Cities for Optimized Urban Development and Planning* (pp. 238–253). IGI Global. doi:10.4018/978-1-5225-3613-0.ch013

Bąkowska-Waldmann, E., & Kaczmarek, T. (2019). The Use of Geo-Questionnaire in Spatial Planning. *International Journal of E-Planning Research*, *8*(2), 45–67. doi:10.4018/IJEPR.2019040103

Bala, I., & Singh, G. (2019). Green Communication for Cognitive Cities. In K. Ahuja & A. Khosla (Eds.), *Driving the Development, Management, and Sustainability of Cognitive Cities* (pp. 87–110). IGI Global. doi:10.4018/978-1-5225-8085-0.ch004

Bañgate, J. M., Dugdale, J., Beck, E., & Adam, C. (2019). Review of Agent Based Modelling of Social Attachment in Crisis Situations. *International Journal of Information Systems for Crisis Response and Management*, *11*(1), 35–64. doi:10.4018/IJISCRAM.2019010103

Bangui, H., Ge, M., & Buhnova, B. (2019). A Research Roadmap of Big Data Clustering Algorithms for Future Internet of Things. *International Journal of Organizational and Collective Intelligence*, *9*(2), 16–30. doi:10.4018/IJOCI.2019040102

Bansal, A., Ahirwar, M. K., & Shukla, P. K. (2019). Assessment on Different Classification Algorithms Used in Internet of Things Applications. *International Journal of Organizational and Collective Intelligence*, *9*(1), 1–11. doi:10.4018/IJOCI.2019010101

Bansal, A., Shukla, P. K., & Ahirwar, M. K. (2019). Opinion on Different Classification Algorithms Used in Internet of Things Environment for Large Data Set. *International Journal of Organizational and Collective Intelligence*, *9*(1), 51–60. doi:10.4018/IJOCI.2019010104

Bansal, R., & Singh, V. K. (2020). Proposed Technique for Efficient Cloud Computing Model in Effective Digital Training Towards Sustainable Livelihoods for Unemployed Youths. *International Journal of Cloud Applications and Computing*, *10*(4), 13–27. doi:10.4018/IJCAC.2020100102

Baradaran, V., Farokhi, S., & Ahamdi, Z. (2018). A Model for Evaluation and Development of Citizens' Electronic Readiness for Deployment of an E-City Using Structural Equation Modeling. *Journal of Global Information Management*, *26*(4), 135–157. doi:10.4018/JGIM.2018100108

Barbery-Montoya, D. C., Del Hierro, C. A., & Arroba-Hurtado, N. Y. (2020). System and Environment for Tourism 4.0. In J. M. Palma-Ruiz, J. Saiz-Álvarez, & Á. Herrero-Crespo (Eds.), *Handbook of Research on Smart Territories and Entrepreneurial Ecosystems for Social Innovation and Sustainable Growth* (pp. 152–171). IGI Global. doi:10.4018/978-1-7998-2097-0.ch009

Baron, M., & Saryazdi, A. K. (2019). Public and Public-Private Cooperation in Building Resilient Urban Logistics. In J. Gonzalez-Feliu, M. Chong, J. Vargas Florez, & J. Padilla Solis (Eds.), *Handbook of Research on Urban and Humanitarian Logistics* (pp. 1–21). IGI Global. doi:10.4018/978-1-5225-8160-4.ch001

Batsaris, M., Kavroudakis, D., Soulakellis, N. A., & Kontos, T. (2019). Location-Allocation Modeling for Emergency Evacuation Planning in a Smart City Context. *International Journal of Applied Geospatial Research*, *10*(4), 28–43. doi:10.4018/IJAGR.2019100103

Belay, D. G. (2019). Positioning Agency of Migrant Street Working Children in Addis Ababa, Ethiopia. *International Journal of Political Activism and Engagement*, *6*(2), 42–59. doi:10.4018/IJPAE.2019040104

Belay, D. G., & Robi, G. A. (2018). Socio-Economic and Psychological Risks of Unemployed Youth in Developing Countries. *International Journal of Risk and Contingency Management*, *7*(2), 67–82. doi:10.4018/IJRCM.2018040104

Belhocine, O., Djiar, K. A., & Lagati, M. (2019). Performance Evaluation of the National Housing File (FNL) for the Development of E-Governance in the Housing Sector in Algeria. *International Journal of E-Planning Research*, *8*(4), 60–73. doi:10.4018/IJEPR.2019100104

Belkeziz, R., & Jarir, Z. (2020). An Overview of the IoT Coordination Challenge. *International Journal of Service Science, Management, Engineering, and Technology*, *11*(1), 99–115. doi:10.4018/IJSSMET.2020010107

Bellavista, P., Corradi, A., Foschini, L., Gomes, E. H., Lamberti, E., Klein, G., De Rolt, C. R., & Torello, M. (2020). Virtual Environments as Enablers of Civic Awareness and Engagement. *International Journal of Urban Planning and Smart Cities*, *1*(1), 22–34. doi:10.4018/IJUPSC.2020010102

Belli, B., & Koramaz, T. K. (2020). Transformation of Urban Space by Smart Technologies. In H. Abusaada, A. M. Salama, & A. Elshater (Eds.), *Reconstructing Urban Ambiance in Smart Public Places* (pp. 82–113). IGI Global. doi:10.4018/978-1-7998-3856-2.ch005

Bena, R. A. (2019). The Role of Industrial Development in Nigeria's Transformation From Oil to Cities. In U. G. Benna (Ed.), *Industrial and Urban Growth Policies at the Sub-National, National, and Global Levels* (pp. 320–341). IGI Global. doi:10.4018/978-1-5225-7625-9.ch016

Benna, I. U. (2018). Improving Quality of Urban Life Through Enhanced Energy Policy in Africa. In U. G. Benna & A. U. Benna (Eds.), *Crowdfunding and Sustainable Urban Development in Emerging Economies* (pp. 255–276). IGI Global. doi:10.4018/978-1-5225-3952-0.ch013

Benna, U. G. (2019). A New Colorful Pathway Chosen. In U. G. Benna (Ed.), *Industrial and Urban Growth Policies at the Sub-National, National, and Global Levels* (pp. 1–25). IGI Global. doi:10.4018/978-1-5225-7625-9.ch001

Benna, U. G. (2019). Transformative Entrepreneurs and Urban Development Deficit in Africa. In U. G. Benna (Ed.), *Optimizing Regional Development Through Transformative Urbanization* (pp. 94–117). IGI Global. doi:10.4018/978-1-5225-5448-6.ch005

Bernardino, S., & Santos, J. F. (2020). Integration Challenges of Immigrants in Smart Cities. *International Journal of Entrepreneurship and Governance in Cognitive Cities, 1*(2), 39–56. doi:10.4018/IJEGCC.2020070104

Beutelspacher, L., Mainka, A., & Siebenlist, T. (2018). Citizen Participation via Mobile Applications. *International Journal of Electronic Government Research, 14*(4), 18–26. doi:10.4018/IJEGR.2018100102

Bhatia, P., & Singh, P. (2019). Technological and Gamified Solutions for Pollution Control in Cognitive Cities. In K. Ahuja & A. Khosla (Eds.), *Driving the Development, Management, and Sustainability of Cognitive Cities* (pp. 234–249). IGI Global. doi:10.4018/978-1-5225-8085-0.ch010

Biao, I. (2019). Urban Planning and the Creation of Livelihoods in Sub-Saharan Africa in the 21st Century. In I. Biao (Ed.), *Learning Cities, Town Planning, and the Creation of Livelihoods* (pp. 55–71). IGI Global. doi:10.4018/978-1-5225-8134-5.ch003

Bigi, D. (2019). The Pillar Porticus in the Architectural Design of High Imperial Epoch. In C. Inglese & A. Ippolito (Eds.), *Conservation, Restoration, and Analysis of Architectural and Archaeological Heritage* (pp. 256–285). IGI Global. doi:10.4018/978-1-5225-7555-9.ch011

Bild, E., Steele, D., Pfeffer, K., Bertolini, L., & Guastavino, C. (2018). Activity as a Mediator Between Users and Their Auditory Environment in an Urban Pocket Park. In F. Aletta & J. Xiao (Eds.), *Handbook of Research on Perception-Driven Approaches to Urban Assessment and Design* (pp. 100–125). IGI Global. doi:10.4018/978-1-5225-3637-6.ch005

Biswas, A., & Maurya, K. K. (2018). Governance and Institutional Framework for Smart Cities in India. In J. Mugambwa & M. W. Katusiimeh (Eds.), *Handbook of Research on Urban Governance and Management in the Developing World* (pp. 50–83). IGI Global. doi:10.4018/978-1-5225-4165-3.ch004

Bolívar, M. P. (2018). Governance in Smart Cities: A Comparison of Practitioners' Perceptions and Prior Research. *International Journal of E-Planning Research, 7*(2), 1–19. doi:10.4018/IJEPR.2018040101

Bolívar, M. P. (2019). Governance in Smart Cities. In Smart Cities and Smart Spaces (pp. 1688-1707). IGI Global. doi:10.4018/978-1-5225-7030-1.ch076

Bottaccioli, L., Patti, E., Osello, A., Cerquitelli, T., Macii, E., & Acquaviva, A. (2019). IoT Platforms and Technologies Driving Spatial Planning and Analytics. In A. Voghera & L. La Riccia (Eds.), *Spatial Planning in the Big Data Revolution* (pp. 84–112). IGI Global. doi:10.4018/978-1-5225-7927-4.ch004

Boumahdi, M., El Amrani, C., & Denys, S. (2019). An Innovative Air Purification Method and Neural Network Algorithm Applied to Urban Streets. *International Journal of Embedded and Real-Time Communication Systems*, *10*(4), 1–19. doi:10.4018/IJERTCS.2019100101

Brown, S., & Bousalis, R. (2018). *Best Practices in K-12 Social Studies Integration*. IGI Global. doi:10.4018/978-1-5225-4065-6.ch003

Bui, H. T. (2019). Fast Fashion Store Image, Customer Satisfaction and Loyalty on the Young Vietnamese Perspective. *International Journal of Systems and Service-Oriented Engineering*, *9*(1), 1–19. doi:10.4018/IJSSOE.2019010101

Butun, I., & Österberg, P. (2019). Detecting Intrusions in Cyber-Physical Systems of Smart Cities. In R. A. Shaikh (Ed.), *Secure Cyber-Physical Systems for Smart Cities* (pp. 74–102). IGI Global. doi:10.4018/978-1-5225-7189-6.ch004

Buyana, K., Lwasa, S., & Kasaija, P. (2019). Gender Ideologies and Climate Risk. *International Journal of Social Ecology and Sustainable Development*, *10*(1), 16–30. doi:10.4018/IJSESD.2019010102

Buyana, K., & Nadiope, M. J. (2019). Thinking Across Sectors. *International Journal of Public and Private Perspectives on Healthcare, Culture, and the Environment*, *3*(1), 1–13. doi:10.4018/IJPPPHCE.2019010101

Bvochora, K., & Kusena, B. (2018). The Political Economy of Land and Urban Development. In J. Mugambwa & M. W. Katusiimeh (Eds.), *Handbook of Research on Urban Governance and Management in the Developing World* (pp. 343–359). IGI Global. doi:10.4018/978-1-5225-4165-3.ch019

Caglayan, N., & Satoglu, S. I. (2019). A Smart Recyclable Waste Collection System Based on IoT and an Application. In Optimizing Big Data Management and Industrial Systems With Intelligent Techniques (pp. 148-169). IGI Global. doi:10.4018/978-1-5225-5137-9.ch007

Calleri, C., Shtrepi, L., Armando, A., & Astolfi, A. (2018). Investigations on the Influence of Auditory Perception on Urban Space Design Through Virtual Acoustics. In F. Aletta & J. Xiao (Eds.), *Handbook of Research on Perception-Driven Approaches to Urban Assessment and Design* (pp. 344–367). IGI Global. doi:10.4018/978-1-5225-3637-6.ch015

Cantarero-García, G. (2020). Intelligence Applied to Smart Cities Through Architecture and Urbanism. In G. Cornetta, A. Touhafi, & G. Muntean (Eds.), *Social, Legal, and Ethical Implications of IoT, Cloud, and Edge Computing Technologies* (pp. 181–202). IGI Global. doi:10.4018/978-1-7998-3817-3.ch008

Cappellaro, F., Chiarini, R., & Meloni, C. (2020). The Humanification of the Urban Community. *International Journal of Urban Planning and Smart Cities*, *1*(1), 35–44. doi:10.4018/IJUPSC.2020010103

Caridi, G. (2018). Calabria 1783. In G. Carlone, N. Martinelli, & F. Rotondo (Eds.), *Designing Grid Cities for Optimized Urban Development and Planning* (pp. 176–187). IGI Global. doi:10.4018/978-1-5225-3613-0.ch010

Carvalho, I., Bidarra, J., & Porto, C. (2018). FeelOpo. *International Journal of Creative Interfaces and Computer Graphics*, *9*(2), 52–62. doi:10.4018/IJCICG.2018070104

Casas, S., Gimeno, J., Casanova-Salas, P., Riera, J. V., & Portalés, C. (2020). Virtual and Augmented Reality for the Visualization of Summarized Information in Smart Cities. In J. M. Rodrigues, P. J. Cardoso, J. Monteiro, & C. M. Ramos (Eds.), *Smart Systems Design, Applications, and Challenges* (pp. 299–325). IGI Global. doi:10.4018/978-1-7998-2112-0.ch015

Çeltek, E., & Ilhan, I. (2020). Big Data, Artificial Intelligence, and Their Implications in the Tourism Industry. In E. Çeltek (Ed.), *Handbook of Research on Smart Technology Applications in the Tourism Industry* (pp. 115–130). IGI Global. doi:10.4018/978-1-7998-1989-9.ch006

Chandiramani, J., & Nayak, S. (2019). Big Data Analytics and Internet of Things for Urban Transportation. In N. Dey & S. Tamane (Eds.), *Big Data Analytics for Smart and Connected Cities* (pp. 244–277). IGI Global. doi:10.4018/978-1-5225-6207-8.ch011

Chang, H. B., & Fang, Y. (2019). Social Design Through Facilitation and Collaboration. In K. Siu & Y. Wong (Eds.), *Practice and Progress in Social Design and Sustainability* (pp. 194–215). IGI Global. doi:10.4018/978-1-5225-4183-7.ch009

Chang, Y., Choi, S. B., Lee, J., & Jin, W. C. (2018). Population Size vs. Number of Crimes: Is the Relationship Superlinear? *International Journal of Information Systems and Social Change*, *9*(1), 26–39. doi:10.4018/IJISSC.2018010102

Charalabidis, Y., & Theocharopoulou, C. (2019). A Participative Method for Prioritizing Smart City Interventions in Medium-Sized Municipalities. *International Journal of Public Administration in the Digital Age*, *6*(1), 41–63. doi:10.4018/IJPADA.2019010103

Chatterjee, S. (2019). Influence of IoT Policy on Quality of Life. *International Journal of Electronic Government Research*, *15*(2), 19–38. doi:10.4018/IJEGR.2019040102

Chaudhary, A., & Tomar, P. (2019). Big Data and IoT Applications in Real Life Environment. In Handbook of Research on Big Data and the IoT (pp. 1-21). IGI Global. doi:10.4018/978-1-5225-7432-3.ch001

Chen, R. (2019). Retail Sector in Singapore. *International Journal of Organizational and Collective Intelligence*, *9*(4), 52–63. doi:10.4018/IJOCI.2019100104

Chen, Y., Zhao, X., Rich, E., & Luna-Reyes, L. F. (2018). Decision Models and Group Decision Support Systems for Emergency Management and City Resilience. *International Journal of E-Planning Research*, *7*(2), 35–50. doi:10.4018/IJEPR.2018040103

Chirisa, I., & Bobo, T. (2018). Informal Sector Operations and the Environment. In J. Mugambwa & M. W. Katusiimeh (Eds.), *Handbook of Research on Urban Governance and Management in the Developing World* (pp. 361–378). IGI Global. doi:10.4018/978-1-5225-4165-3.ch020

Chirisa, I., & Matamanda, A. R. (2019). Smart Cities in Sub-Saharan Africa. In U. G. Benna (Ed.), *Industrial and Urban Growth Policies at the Sub-National, National, and Global Levels* (pp. 83–99). IGI Global. doi:10.4018/978-1-5225-7625-9.ch005

Chirisa, I., Mhlanga, G., Dube, B., & Mukarwi, L. (2019). Metropolitan Councils. In U. G. Benna (Ed.), *Optimizing Regional Development Through Transformative Urbanization* (pp. 252–270). IGI Global. doi:10.4018/978-1-5225-5448-6.ch012

Chirisa, I., Mhlanga, G., & Matamanda, A. R. (2019). Infrastructure, City-Region Development, and Africa's Territorial Spaces. In U. G. Benna (Ed.), *Optimizing Regional Development Through Transformative Urbanization* (pp. 186–203). IGI Global. doi:10.4018/978-1-5225-5448-6.ch009

Chirisa, I., Mukarwi, L., & Matamanda, A. R. (2018). Prospects and Options for Sustainable and Inclusive Crowdfunding in African Cities. In U. G. Benna & A. U. Benna (Eds.), *Crowdfunding and Sustainable Urban Development in Emerging Economies* (pp. 211–231). IGI Global. doi:10.4018/978-1-5225-3952-0.ch011

Chirisa, I., Mukarwi, L., & Matamanda, A. R. (2018). Social Costs and Benefits of the Transformation of the Traditional Families in an African Urban Society. In U. Benna & I. Benna (Eds.), *Urbanization and Its Impact on Socio-Economic Growth in Developing Regions* (pp. 179–197). IGI Global. doi:10.4018/978-1-5225-2659-9.ch009

Chirisa, I., Nyevera, T., & Mandaza-Tsoriyo, W. W. (2019). Industrial and Urban Growth Policies in Sub-Saharan Africa and North Africa Compared. In U. G. Benna (Ed.), *Industrial and Urban Growth Policies at the Sub-National, National, and Global Levels* (pp. 145–167). IGI Global. doi:10.4018/978-1-5225-7625-9.ch008

Chowdhury, A., Karmakar, G., & Kamruzzaman, J. (2019). The Co-Evolution of Cloud and IoT Applications. In S. Singh & R. Mohan Sharma (Eds.), *Handbook of Research on the IoT, Cloud Computing, and Wireless Network Optimization* (pp. 213–234). IGI Global. doi:10.4018/978-1-5225-7335-7.ch011

Chu, S., Tu, S., Wang, N., & Zhang, W. (2020). Information Equity and Cultural Sharing. *International Journal of Library and Information Services*, *9*(1), 10–24. doi:10.4018/IJLIS.2020010102

Coban, G., & Aydın, Ş. (2020). A New Concept in Tourism. In E. Çeltek (Ed.), *Handbook of Research on Smart Technology Applications in the Tourism Industry* (pp. 414–429). IGI Global. doi:10.4018/978-1-7998-1989-9.ch019

Cobbinah, P. B., & Adams, E. A. (2018). Urbanization and Electric Power Crisis in Ghana. In U. Benna & I. Benna (Eds.), *Urbanization and Its Impact on Socio-Economic Growth in Developing Regions* (pp. 262–284). IGI Global. doi:10.4018/978-1-5225-2659-9.ch013

Colic, R., Milovanović Rodić, D., & Maruna, M. (2020). New Concepts and Instruments for the Urban-Rural Continuum. In A. Krstić-Furundžić & A. Djukić (Eds.), *Handbook of Research on Urban-Rural Synergy Development Through Housing, Landscape, and Tourism* (pp. 292–317). IGI Global. doi:10.4018/978-1-5225-9932-6.ch015

Collins, B. K. (2019). Challenges of Isolation for African American Movement to Suburbanization. In P. H. Hoffman-Miller, M. James, & D. Hermond (Eds.), *African American Suburbanization and the Consequential Loss of Identity* (pp. 1–19). IGI Global. doi:10.4018/978-1-5225-7835-2.ch001

Constantinescu, T., Devisch, O., & Kostov, G. (2020). Game Mechanics as Thinking Mechanisms for Urban Development. In C. N. Silva (Ed.), *Citizen-Responsive Urban E-Planning* (pp. 215–251). IGI Global. doi:10.4018/978-1-7998-4018-3.ch009

Constantinescu, T. I., Devisch, O., & Kostov, G. (2017). City Makers. *International Journal of E-Planning Research, 6*(4), 32–57. doi:10.4018/IJEPR.2017100103

Cook, S. B., Cochrane, L., & Corbett, J. (2020). Searching Through Silos. *International Journal of E-Planning Research, 9*(4), 23–39. doi:10.4018/IJEPR.2020100102

Corbett, J., Brennan, S., & Whitely, A. (2020). The Uncertain Participatory Geoweb. In C. N. Silva (Ed.), *Citizen-Responsive Urban E-Planning* (pp. 111–137). IGI Global. doi:10.4018/978-1-7998-4018-3.ch005

Criado, J. I., & Ruvalcaba-Gomez, E. A. (2018). Perceptions of City Managers About Open Government Policies. *International Journal of Electronic Government Research, 14*(1), 1–22. doi:10.4018/IJEGR.2018010101

Cruz-Jesus, F., Oliveira, T., & Bacao, F. (2018). The Global Digital Divide. *Journal of Global Information Management, 26*(2), 1–26. doi:10.4018/JGIM.2018040101

D'Onofrio, S., Habenstein, A., & Portmann, E. (2019). Ontological Design for Cognitive Cities. In K. Ahuja & A. Khosla (Eds.), *Driving the Development, Management, and Sustainability of Cognitive Cities* (pp. 183–211). IGI Global. doi:10.4018/978-1-5225-8085-0.ch008

Dabiri, S., Bakhsh Kelarestaghi, K., & Heaslip, K. (2020). Probe People and Vehicle-Based Data Sources Application in Smart Transportation. In G. Bekdaş, S. M. Nigdeli, & M. Yücel (Eds.), *Artificial Intelligence and Machine Learning Applications in Civil, Mechanical, and Industrial Engineering* (pp. 162–195). IGI Global. doi:10.4018/978-1-7998-0301-0.ch010

Damurski, L., Pluta, J., Ładysz, J., & Mayer-Wydra, M. (2019). Online or Offline Services for Urban Neighbourhoods? *International Journal of E-Planning Research*, *8*(1), 50–67. doi:10.4018/IJEPR.2019010104

Dandotiya, B. (2019). Health Effects of Air Pollution in Urban Environment. In A. Karmaoui (Ed.), *Climate Change and Its Impact on Ecosystem Services and Biodiversity in Arid and Semi-Arid Zones* (pp. 96–115). IGI Global. doi:10.4018/978-1-5225-7387-6.ch006

Das, S., & Nayyar, A. (2019). Innovative Ideas to Manage Urban Traffic Congestion in Cognitive Cities. In K. Ahuja & A. Khosla (Eds.), *Driving the Development, Management, and Sustainability of Cognitive Cities* (pp. 139–162). IGI Global. doi:10.4018/978-1-5225-8085-0.ch006

Davalos, J., & Vargas-Hernández, J. G. (2020). Rescuing Public Spaces for Organic Food Production and Its Economic, Social, and Environmental Impact. *International Journal of Sustainable Economies Management*, *9*(1), 36–51. doi:10.4018/IJSEM.2020010104

De Filippi, F., Coscia, C., & Cocina, G. G. (2020). Digital Participatory Platforms for Urban Regeneration. *International Journal of E-Planning Research*, *9*(3), 47–67. doi:10.4018/IJEPR.2020070103

De Filippi, F., Coscia, C., Cocina, G. G., Lazzari, G., & Manzo, S. (2020). Digital Participatory Platforms for Civic Engagement: A New Way of Participating in Society? *International Journal of Urban Planning and Smart Cities*, *1*(1), 1–21. doi:10.4018/IJUPSC.2020010101

De Filippi, F., Coscia, C., & Guido, R. (2019). From Smart-Cities to Smart-Communities. *International Journal of E-Planning Research*, *8*(2), 24–44. doi:10.4018/IJEPR.2019040102

De Filippi, F., Coscia, C., & Guido, R. (2020). Digital Platforms for Enhancing Participatory Design and Urban Regeneration. In C. N. Silva (Ed.), *Citizen-Responsive Urban E-Planning* (pp. 54–82). IGI Global. doi:10.4018/978-1-7998-4018-3.ch003

De la Calle-Vaquero, M., García-Hernández, M., Mendoza de Miguel, S., & Ferreiro-Calzada, E. (2020). In Search of Overtourism Indicators in Urban Centres. In Handbook of Research on the Impacts, Challenges, and Policy Responses to Overtourism (pp. 302-324). IGI Global. doi:10.4018/978-1-7998-2224-0.ch016

De Matteis, F. (2018). The City as a Mode of Perception. In F. Aletta & J. Xiao (Eds.), *Handbook of Research on Perception-Driven Approaches to Urban Assessment and Design* (pp. 436–457). IGI Global. doi:10.4018/978-1-5225-3637-6.ch018

Dede, G., Fragiadakis, G., Michalakelis, C., & Kamalakis, T. (2019). Brokering Intelligence as a Service for the Internet of Things. *International Journal of Technology Diffusion*, *10*(2), 18–33. doi:10.4018/IJTD.2019040102

Degbelo, A., Wissing, J., & Kauppinen, T. (2018). A Comparison of Geovisualizations and Data Tables for Transparency Enablement in the Open Government Data Landscape. *International Journal of Electronic Government Research*, *14*(4), 39–64. doi:10.4018/IJEGR.2018100104

del Cerro Santamaría, G. (2020). Possibilities and Challenges for Intercultural Research in Global Urbanism. In R. Brunet-Thornton (Ed.), *Examining Cultural Perspectives in a Globalized World* (pp. 94–115). IGI Global. doi:10.4018/978-1-7998-0214-3.ch005

Desarkar, A., Sanyal, S., Baidya, A., Das, A., & Chaudhuri, C. (2019). Innovative Outlier Removal Techniques to Enhance Signature Authentication Accuracy for Smart Society. *International Journal of Distributed Systems and Technologies*, *10*(2), 64–83. doi:10.4018/IJDST.2019040104

Devarakonda, S., Halgamuge, M. N., & Mohammad, A. (2019). Critical Issues in the Invasion of the Internet of Things (IoT). In Handbook of Research on Big Data and the IoT (pp. 174-196). IGI Global. doi:10.4018/978-1-5225-7432-3.ch010

Devkota, K. (2018). Challenges of Inclusive Urbanization in the Face of Political Transition in Nepal. In J. Mugambwa & M. W. Katusiimeh (Eds.), *Handbook of Research on Urban Governance and Management in the Developing World* (pp. 159–171). IGI Global. doi:10.4018/978-1-5225-4165-3.ch009

Dhage, V. (2020). The Gastronomy of the Eye. In H. Abusaada, A. M. Salama, & A. Elshater (Eds.), *Reconstructing Urban Ambiance in Smart Public Places* (pp. 175–198). IGI Global. doi:10.4018/978-1-7998-3856-2.ch009

Dinić-Branković, M., Đekić, J., Mitković, P., & Igić, M. (2020). Restructuring Centers of Former Rural Settlements and Weekend-Residence Zones. In A. Krstić-Furundžić & A. Djukić (Eds.), *Handbook of Research on Urban-Rural Synergy Development Through Housing, Landscape, and Tourism* (pp. 361–379). IGI Global. doi:10.4018/978-1-5225-9932-6.ch018

Dionisio, R. D., Schindler, M., & Kingham, S. (2020). Tools for Sustainable Change. *International Journal of E-Planning Research*, *9*(2), 21–42. doi:10.4018/IJEPR.2020040102

Dixit, S. (2020). Impact of Urban Ambiance on the Productivity of the Built Environment Using Cluster Analysis. In H. Abusaada, A. M. Salama, & A. Elshater (Eds.), *Reconstructing Urban Ambiance in Smart Public Places* (pp. 126–144). IGI Global. doi:10.4018/978-1-7998-3856-2.ch007

Djukic, A., Stupar, A., & Antonic, B. M. (2018). The Orthogonal Urban Matrix of the Towns in Vojvodina, Northern Serbia. In G. Carlone, N. Martinelli, & F. Rotondo (Eds.), *Designing Grid Cities for Optimized Urban Development and Planning* (pp. 128–156). IGI Global. doi:10.4018/978-1-5225-3613-0.ch008

Djukić, A. M., & Antonić, B. M. (2020). Trends in New Multi-Family Residential Development in Serbia. In A. Krstić-Furundžić & A. Djukić (Eds.), *Handbook of Research on Urban-Rural Synergy Development Through Housing, Landscape, and Tourism* (pp. 30–50). IGI Global. doi:10.4018/978-1-5225-9932-6.ch002

Dobrowolski, J., Kobylarczyk, J., Wagner, A. L., & Mazur, R. (2018). Involving Diverse Stakeholders for Sustainable Development. In U. C. Pandey & V. Indrakanti (Eds.), *Optimizing Open and Distance Learning in Higher Education Institutions* (pp. 336–358). IGI Global. doi:10.4018/978-1-5225-2624-7.ch016

Doctor, G., & Patel, P. (2019). Air Quality Monitoring Using Internet of Things (IoT) in Smart Cities. In K. Saravanan, G. Julie, & H. Robinson (Eds.), *Handbook of Research on Implementation and Deployment of IoT Projects in Smart Cities* (pp. 317–333). IGI Global. doi:10.4018/978-1-5225-9199-3.ch019

Dodo, O., & Mwale, C. (2018). Rural Urban Youth Delinquency in Zimbabwe, the Nexus. In J. Mugambwa & M. W. Katusiimeh (Eds.), *Handbook of Research on Urban Governance and Management in the Developing World* (pp. 207–222). IGI Global. doi:10.4018/978-1-5225-4165-3.ch012

Dogan, O., & Gurcan, O. F. (2019). Applications of Big Data and Green IoT-Enabling Technologies for Smart Cities. In Handbook of Research on Big Data and the IoT (pp. 22-41). IGI Global. doi:10.4018/978-1-5225-7432-3.ch002

Doheim, R. M., Farag, A. A., & Badawi, S. (2020). Success Measures for Transforming Into Car-Free Cities. In R. M. Doheim, A. A. Farag, & E. Kamel (Eds.), *Humanizing Cities Through Car-Free City Development and Transformation* (pp. 231–267). IGI Global. doi:10.4018/978-1-7998-3507-3.ch010

Dorrell, D. (2018). Using International Content in an Introductory Human Geography Course. In S. Dikli, B. Etheridge, & R. Rawls (Eds.), *Curriculum Internationalization and the Future of Education* (pp. 55–80). IGI Global. doi:10.4018/978-1-5225-2791-6.ch004

Dorrell, D. (2020). Using OpenStreetMap as a Source of Religious Mapping Data in an OER Human Geography Textbook. In M. Y. Zhou (Ed.), *Open Educational Resources (OER) Pedagogy and Practices* (pp. 112–139). IGI Global. doi:10.4018/978-1-7998-1200-5.ch006

Doyle, A., Hynes, W., & Purcell, S. M. (2020). Enhancing Urban Resilience. In C. N. Silva (Ed.), *Citizen-Responsive Urban E-Planning* (pp. 282–304). IGI Global. doi:10.4018/978-1-7998-4018-3.ch011

Dutta Roy, A., Debnath Parial, M., & Mukherjee, K. (2020). Synergy Between Air Quality, Various Urban Forms, and Land Surface Temperature. In A. C. Affam & E. H. Ezechi (Eds.), *Handbook of Research on Resource Management for Pollution and Waste Treatment* (pp. 576–609). IGI Global. doi:10.4018/978-1-7998-0369-0.ch024

Dyson, J. (2020). Highway Robbery. In R. M. Doheim, A. A. Farag, & E. Kamel (Eds.), *Humanizing Cities Through Car-Free City Development and Transformation* (pp. 204–230). IGI Global. doi:10.4018/978-1-7998-3507-3.ch009

Efiong, J. E., & Adegbola, A. S. (2020). Leveraging on Digital Technologies to Up-Scale Tourism for Economic Growth in Africa. *International Journal of Tourism and Hospitality Management in the Digital Age, 4*(1), 42–54. doi:10.4018/IJTHMDA.2020010104

Eilu, E. (2018). An Assessment of Mobile Internet Usage in a Rural Setting of a Developing Country. *International Journal of Mobile Computing and Multimedia Communications, 9*(2), 47–59. doi:10.4018/IJMCMC.2018040104

Ekman, U. (2018). Smart City Planning. *International Journal of E-Planning Research, 7*(3), 1–21. doi:10.4018/IJEPR.2018070101

El-Kholei, A. O., & Al-Jayyousi, O. (2018). Cities of GCC in Post-Hydrocarbon Era. In A. A. Alraouf (Ed.), *Knowledge-Based Urban Development in the Middle East* (pp. 248–270). IGI Global. doi:10.4018/978-1-5225-3734-2.ch013

El Sayed, O. (2021). Knowledge-Based Policymaking for Urban Development. In A. A. Galaby & A. A. Abdrabo (Eds.), *Handbook of Research on Creative Cities and Advanced Models for Knowledge-Based Urban Development* (pp. 240–264). IGI Global. doi:10.4018/978-1-7998-4948-3.ch012

Elgarej, M., Mansouri, K., & Youssfi, M. (2020). Modelling of Logistics Monitoring System for Milk Collection Based on Swarm Intelligence. *International Journal of Smart Vehicles and Smart Transportation, 3*(2), 59–82. doi:10.4018/IJSVST.2020070104

Elgobashi, A., & Elsemary, Y. M. (2020). Theorizing the Critical Placemaking. In H. Abusaada, A. M. Salama, & A. Elshater (Eds.), *Reconstructing Urban Ambiance in Smart Public Places* (pp. 145–174). IGI Global. doi:10.4018/978-1-7998-3856-2.ch008

Elm, J., & Carvalho, L. C. (2020). Best Practices to Become a Sustainable Smart City. In I. S. Paiva & L. C. Carvalho (Eds.), *Conceptual and Theoretical Approaches to Corporate Social Responsibility, Entrepreneurial Orientation, and Financial Performance* (pp. 247–265). IGI Global. doi:10.4018/978-1-7998-2128-1.ch013

Elsantil, Y. (2020). User Perceptions of the Security of Mobile Applications. *International Journal of E-Services and Mobile Applications*, *12*(4), 24–41. doi:10.4018/IJESMA.2020100102

Endong, F. P., & Essoh, G. E. (2020). Ode to Africa's Urban Space. In F. P. Endong (Ed.), *Deconstructing Images of the Global South Through Media Representations and Communication* (pp. 166–186). IGI Global. doi:10.4018/978-1-5225-9821-3.ch006

Ergen, B., & Oğuz, M. Ö. (2019). Researching Sustainable Rural Development With Location Quotient Method in Yozgat (2006-2016). In Q. R. Yasser (Ed.), *Handbook of Research on Rural Sociology and Community Mobilization for Sustainable Growth* (pp. 179–207). IGI Global. doi:10.4018/978-1-5225-7158-2.ch011

Eweida, A. M. (2021). Urban Laws in Harmony or at Odds With Knowledge-Based Urban Policies. In A. A. Galaby & A. A. Abdrabo (Eds.), *Handbook of Research on Creative Cities and Advanced Models for Knowledge-Based Urban Development* (pp. 23–40). IGI Global. doi:10.4018/978-1-7998-4948-3.ch002

Eybalin Casseus, C., Odongoh, S., & Abdrabo, A. (2021). *Pushing the Limits of Mediterranean Configurations*. IGI Global. doi:10.4018/978-1-7998-4438-9.ch001

Eynard, E., Melis, G., & Tabasso, M. (2020). Walkable Urban Environments for Wellbeing. *International Journal of Urban Planning and Smart Cities*, *1*(2), 17–29. doi:10.4018/IJUPSC.2020070102

Falco, E., Stylianou, C., Martinez, G., Kleinhans, R., Basso-Moro, S., & Neophytou, H. (2020). User Acceptance of Technology. *International Journal of Electronic Government Research*, *16*(3), 85–104. doi:10.4018/IJEGR.2020070105

Fang, D., Ye, F., Qian, Y., & Sharif, H. (2017). A Cloud-Based Incentive Mechanism for Sensing in Mobile Sensor Networks. *International Journal of Handheld Computing Research*, *8*(3), 1–14. doi:10.4018/IJHCR.2017070101

Farag, A. A., Badawi, S., & Doheim, R. M. (2019). Digital Simulation for the Outdoor Thermal Comfort Assessment. In H. Abusaada, C. Vellguth, & A. Elshater (Eds.), *Handbook of Research on Digital Research Methods and Architectural Tools in Urban Planning and Design* (pp. 33–46). IGI Global. doi:10.4018/978-1-5225-9238-9.ch003

Farinmade, A. A., Soyinka, O. A., & Siu, K. W. (2018). Urban Safety and Security in Lagos Metropolis, Nigeria. In J. Mugambwa & M. W. Katusiimeh (Eds.), *Handbook of Research on Urban Governance and Management in the Developing World* (pp. 193–206). IGI Global. doi:10.4018/978-1-5225-4165-3.ch011

Fasolino, I. (2018). Rules for a New Town After a Disaster. In G. Carlone, N. Martinelli, & F. Rotondo (Eds.), *Designing Grid Cities for Optimized Urban Development and Planning* (pp. 157–175). IGI Global. doi:10.4018/978-1-5225-3613-0.ch009

Fazzin, S. (2020). *The Fourth Industrial Revolution and the Internet of Things*. IGI Global. doi:10.4018/978-1-7998-2036-9.ch001

Flores, C. A. (2020). Qualitative Analysis of Learning Territorial Planning. In J. M. Palma-Ruiz, J. Saiz-Álvarez, & Á. Herrero-Crespo (Eds.), *Handbook of Research on Smart Territories and Entrepreneurial Ecosystems for Social Innovation and Sustainable Growth* (pp. 213–232). IGI Global. doi:10.4018/978-1-7998-2097-0.ch012

Florin, R., & Olariu, S. (2018). Vehicular Clouds. In J. Grover, P. Vinod, & C. Lal (Eds.), *Vehicular Cloud Computing for Traffic Management and Systems* (pp. 1–29). IGI Global. doi:10.4018/978-1-5225-3981-0.ch001

Frank, E., & Aznar Fernández-Montesinos, G. (2020). Smart City = Smart Citizen = Smart Economy? In G. Cornetta, A. Touhafi, & G. Muntean (Eds.), *Social, Legal, and Ethical Implications of IoT, Cloud, and Edge Computing Technologies* (pp. 161–180). IGI Global. doi:10.4018/978-1-7998-3817-3.ch007

Fredericks, J., Tomitsch, M., & Haeusler, M. H. (2020). Redefining Community Engagement in Smart Cities. In C. N. Silva (Ed.), *Citizen-Responsive Urban E-Planning* (pp. 13–53). IGI Global. doi:10.4018/978-1-7998-4018-3.ch002

Frizziero, L., Donnici, G., Caligiana, G., Liverani, A., & Francia, D. (2018). Project of Inventive Ideas Through a TRIZ Study Applied to the Analysis of an Innovative Urban Transport Means. *International Journal of Manufacturing, Materials, and Mechanical Engineering*, *8*(4), 35–62. doi:10.4018/IJMMME.2018100103

Furundžić, N. Z., Furundžić, D. P., & Krstić-Furundžić, A. (2020). Urban-Rural Synergy Toward Human Well-Being. In A. Krstić-Furundžić & A. Djukić (Eds.), *Handbook of Research on Urban-Rural Synergy Development Through Housing, Landscape, and Tourism* (pp. 51–70). IGI Global. doi:10.4018/978-1-5225-9932-6.ch003

Galal, O. M. (2019). An Argumentative Review Between Urban Morphology and Energy-Efficient Mobility. In H. Abusaada, C. Vellguth, & A. Elshater (Eds.), *Handbook of Research on Digital Research Methods and Architectural Tools in Urban Planning and Design* (pp. 283–309). IGI Global. doi:10.4018/978-1-5225-9238-9.ch013

Gambarato, R. R. (2018). Transmedia Journalism and the City. In R. R. Gambarato & G. C. Alzamora (Eds.), *Exploring Transmedia Journalism in the Digital Age* (pp. 147–161). IGI Global. doi:10.4018/978-1-5225-3781-6.ch009

Gan, L., & Zhang, W. (2021). Infrastructure and Industry Economy. In K. Lyu, M. Hu, J. Du, & V. Sugumaran (Eds.), *AI-Based Services for Smart Cities and Urban Infrastructure* (pp. 118–139). IGI Global. doi:10.4018/978-1-7998-5024-3.ch006

Gao, X., & Gan, L. (2021). Framework and Structure of Smart Cities. In K. Lyu, M. Hu, J. Du, & V. Sugumaran (Eds.), *AI-Based Services for Smart Cities and Urban Infrastructure* (pp. 102–116). IGI Global. doi:10.4018/978-1-7998-5024-3.ch005

Garcia, B. C. (2019). Cities, Knowledge Flows and the New Commons. In U. G. Benna (Ed.), *Optimizing Regional Development Through Transformative Urbanization* (pp. 71–92). IGI Global. doi:10.4018/978-1-5225-5448-6.ch004

Garcia, B. C. (2019). Green and Blue City-Regions. In U. G. Benna (Ed.), *Industrial and Urban Growth Policies at the Sub-National, National, and Global Levels* (pp. 169–190). IGI Global. doi:10.4018/978-1-5225-7625-9.ch009

Garcia, I., Herranz-Pascual, K., Aspuru, I., Gutierrez, L., Acero, J. A., & Santander, A. (2018). Indicators of Environmental Comfort Sensitive to Human Perception. In F. Aletta & J. Xiao (Eds.), *Handbook of Research on Perception-Driven Approaches to Urban Assessment and Design* (pp. 508–533). IGI Global. doi:10.4018/978-1-5225-3637-6.ch022

Garzilli, F., Mazzarella, C., & Vittiglio, V. (2020). Integrated Approaches for Peri-Urban Wastescapes. *International Journal of Urban Planning and Smart Cities*, *1*(2), 43–58. doi:10.4018/IJUPSC.2020070104

Garzino, G., Novello, G., & Bocconcino, M. M. (2019). Handbook of Research on Urban and Territorial Systems and the Intangible Dimension. In C. Inglese & A. Ippolito (Eds.), *Conservation, Restoration, and Analysis of Architectural and Archaeological Heritage* (pp. 346–385). IGI Global. doi:10.4018/978-1-5225-7555-9.ch014

Gharehbaghi, K., Farnes, K., & Myers, M. (2020). Melbourne's Advanced Rail Transportation. *International Journal of Strategic Engineering*, *3*(2), 24–36. doi:10.4018/IJoSE.2020070102

Ghoni, R., & Ibrahim, T. (2018). Internet of Things-Based Emergency Response Management. *International Journal of Disaster Response and Emergency Management*, *1*(1), 38–50. doi:10.4018/IJDREM.2018010103

Gil, O., Cortés-Cediel, M. E., & Cantador, I. (2019). Citizen Participation and the Rise of Digital Media Platforms in Smart Governance and Smart Cities. *International Journal of E-Planning Research*, *8*(1), 19–34. doi:10.4018/IJEPR.2019010102

Göktaş Kulualp, H., & Sarı, Ö. (2020). Smart Tourism, Smart Cities, and Smart Destinations as Knowledge Management Tools. In E. Çeltek (Ed.), *Handbook of Research on Smart Technology Applications in the Tourism Industry* (pp. 371–390). IGI Global. doi:10.4018/978-1-7998-1989-9.ch017

Gölbey, A. G., & Kaylı, A. (2021). Urbanization Violence to Nature. In M. N. Erdem, N. Kocabay-Sener, & T. Demir (Eds.), *Handbook of Research on Aestheticization of Violence, Horror, and Power* (pp. 141–155). IGI Global. doi:10.4018/978-1-7998-4655-0.ch008

Golden Julie, E. (2019). Waste Management System for Smart City Using IoT. In D. Goyal, S. Balamurugan, S. Peng, & D. S. Jat (Eds.), *The IoT and the Next Revolutions Automating the World* (pp. 1–15). IGI Global. doi:10.4018/978-1-5225-9246-4.ch001

Gomes, C. C. (2020). From Smart Concept to Desired Experiences. In L. Farinha & D. Raposo (Eds.), *Handbook of Research on Driving Industrial Competitiveness With Innovative Design Principles* (pp. 154–168). IGI Global. doi:10.4018/978-1-7998-3628-5.ch011

Gonçalves, C. (2018). Models, Methods, and Metrics to Measure Socioeconomic Resilience. In I. D. Rosa, J. C. Lopes, R. Ribeiro, & A. Mendes (Eds.), *Handbook of Research on Methods and Tools for Assessing Cultural Landscape Adaptation* (pp. 346–367). IGI Global. doi:10.4018/978-1-5225-4186-8.ch014

Gonçalves, C., & Claudino, S. (2021). Governance, Civic and Geographic Education, and Local Development. In Teaching and Learning Practices That Promote Sustainable Development and Active Citizenship (pp. 236-253). IGI Global. doi:10.4018/978-1-7998-4402-0.ch012

González, E. H., & Monnet, J. (2018). Walking as Kinaesthetic Experience of the City. In F. Aletta & J. Xiao (Eds.), *Handbook of Research on Perception-Driven Approaches to Urban Assessment and Design* (pp. 477–488). IGI Global. doi:10.4018/978-1-5225-3637-6.ch020

Goodchild, B. (2020). Conceptualising the Use of Digital Technologies in Spatial Planning. *International Journal of E-Planning Research*, *9*(3), 1–23. doi:10.4018/IJEPR.2020070101

Goodspeed, R. (2019). Participatory E-Planning With Civic Crowdfunding. *International Journal of E-Planning Research*, *8*(2), 68–87. doi:10.4018/IJEPR.2019040104

Gordin, V., Borovskaia, I., & Fedorova, E. (2020). Overtourism in Megalopolises. In Handbook of Research on the Impacts, Challenges, and Policy Responses to Overtourism (pp. 116-135). IGI Global. doi:10.4018/978-1-7998-2224-0.ch007

Goyal, P., & Singh, P. (2020). Improved Design of Vertical Cavity Surface Emitting Laser for 3D Sensing in Internet of Things Applications. In A. Solanki & A. Nayyar (Eds.), *Green Building Management and Smart Automation* (pp. 64–89). IGI Global. doi:10.4018/978-1-5225-9754-4.ch003

Graziano, T. (2020). Rural Entrepreneurship, Innovation, and Technology. In A. Jean Vasile, J. Subic, A. Grubor, & D. Privitera (Eds.), *Handbook of Research on Agricultural Policy, Rural Development, and Entrepreneurship in Contemporary Economies* (pp. 334–353). IGI Global. doi:10.4018/978-1-5225-9837-4.ch017

Greene, M., Clarke-Hagan, D., & Curran, M. (2020). Achieving Smarter Buildings and More Efficient Facilities Management. *International Journal of Digital Innovation in the Built Environment*, *9*(2), 1–16. doi:10.4018/IJDIBE.2020070101

Grochulska-Salak, M. (2019). Urban Farming in Sustainable City Development. In J. G. Vargas-Hernández & J. A. Zdunek-Wielgołaska (Eds.), *Bioeconomical Solutions and Investments in Sustainable City Development* (pp. 43–64). IGI Global. doi:10.4018/978-1-5225-7958-8.ch003

Guhathakurta, S., Zhang, G., Chen, G., Burnette, C., & Sepkowitz, I. (2019). Mining Social Media to Measure Neighborhood Quality in the City of Atlanta. *International Journal of E-Planning Research*, *8*(1), 1–18. doi:10.4018/IJEPR.2019010101

Guma, I. P., Rwashana, A. S., & Oyo, B. (2018). Food Security Indicators for Subsistence Farmers Sustainability. *International Journal of System Dynamics Applications*, *7*(1), 45–64. doi:10.4018/IJSDA.2018010103

Gunalay, Y., & Yeomans, J. S. (2020). An Innovative Modelling and Decision-Support Approach for Evaluating Urban Transshipment Problems Using Electrical Trucks. *International Journal of Smart Vehicles and Smart Transportation*, *3*(2), 19–37. doi:10.4018/IJSVST.2020070102

Gupta, B. B., Chaudhary, P., & Gupta, S. (2020). Designing a XSS Defensive Framework for Web Servers Deployed in the Existing Smart City Infrastructure. *Journal of Organizational and End User Computing*, *32*(4), 85–111. doi:10.4018/JOEUC.2020100105

Gupta, P., & Jain, V. K. (2020). Interpretive Structural Modeling of GIoT enablers. *Journal of Information Technology Research*, *13*(2), 129–140. doi:10.4018/JITR.2020040108

Gupta, R. (2019). Resource Provisioning and Scheduling Techniques of IoT Based Applications in Fog Computing. *International Journal of Fog Computing*, *2*(2), 57–70. doi:10.4018/IJFC.2019070104

Gurney, R. H. III, Alomaim, A., & Al-Bader, J. (2018). Kuwait Urban App. In A. A. Alraouf (Ed.), *Knowledge-Based Urban Development in the Middle East* (pp. 186–203). IGI Global. doi:10.4018/978-1-5225-3734-2.ch010

Gutiérrez-Diez, M. D., Bordas-Beltrán, J. L., & Arras-Vota, A. M. (2020). Smart Territory Initiatives in an Emerging Economy. In J. M. Palma-Ruiz, J. Saiz-Álvarez, & Á. Herrero-Crespo (Eds.), *Handbook of Research on Smart Territories and Entrepreneurial Ecosystems for Social Innovation and Sustainable Growth* (pp. 191–212). IGI Global. doi:10.4018/978-1-7998-2097-0.ch011

Gutiérrez-Leefmans, M. (2020). The Role of Business in the Innovation Ecosystem. In J. M. Palma-Ruiz, J. Saiz-Álvarez, & Á. Herrero-Crespo (Eds.), *Handbook of Research on Smart Territories and Entrepreneurial Ecosystems for Social Innovation and Sustainable Growth* (pp. 19–36). IGI Global. doi:10.4018/978-1-7998-2097-0.ch002

Hahn, J. (2020). Student Engagement and Smart Spaces. In B. Holland (Ed.), *Emerging Trends and Impacts of the Internet of Things in Libraries* (pp. 52–70). IGI Global. doi:10.4018/978-1-7998-4742-7.ch003

Hai-Jew, S. (2019). *Ferreting Out Silences and Invisibilities in Electronic Hive Minds*. IGI Global. doi:10.4018/978-1-5225-9369-0.ch002

Hakky, R. I. (2020). Potentials of ICT and IoT in the Design of Smart Public Spaces. In H. Abusaada, A. M. Salama, & A. Elshater (Eds.), *Reconstructing Urban Ambiance in Smart Public Places* (pp. 60–81). IGI Global. doi:10.4018/978-1-7998-3856-2.ch004

Hamid, B., Jhanjhi, N. Z., & Humayun, M. (2020). Digital Governance for Developing Countries Opportunities, Issues, and Challenges in Pakistan. In V. Ponnusamy, K. Rafique, & N. Zaman (Eds.), *Employing Recent Technologies for Improved Digital Governance* (pp. 36–58). IGI Global. doi:10.4018/978-1-7998-1851-9.ch003

Haq, F., & Medhekar, A. (2018). The Economic Effects of Spiritual Tourism in India and Pakistan. In U. Benna & I. Benna (Eds.), *Urbanization and Its Impact on Socio-Economic Growth in Developing Regions* (pp. 66–86). IGI Global. doi:10.4018/978-1-5225-2659-9.ch004

Henriques, C. H. (2020). Lisbon Historic Quarters. In Handbook of Research on the Impacts, Challenges, and Policy Responses to Overtourism (pp. 285-301). IGI Global. doi:10.4018/978-1-7998-2224-0.ch015

Henshilwood, E., Swilling, M., & Naidoo, M. L. (2019). A Transdisciplinary Inquiry Into Sustainable Automobility Transitions. *International Journal of E-Planning Research*, *8*(3), 13–37. doi:10.4018/IJEPR.2019070102

Hepburn, P. A. (2018). A New Governance Model for Delivering Digital Policy Agendas. *International Journal of E-Planning Research*, *7*(3), 36–49. doi:10.4018/IJEPR.2018070103

Hipps, R., Chopra, T., Zhao, P., Kwartler, E., & Jaume, S. (2017). Geospatial Analytics to Improve the Safety of Autonomous Vehicles. *International Journal of Knowledge-Based Organizations*, *7*(3), 40–51. doi:10.4018/IJKBO.2017070104

Holland, B. (2020). Emerging Technology and Today's Libraries. In B. Holland (Ed.), *Emerging Trends and Impacts of the Internet of Things in Libraries* (pp. 1–33). IGI Global. doi:10.4018/978-1-7998-4742-7.ch001

Honarvar, A. R., & Sami, A. (2018). Improve Home Energy Management System by Extracting Usage Patterns From Power Usage Big Data of Homes' Appliances. In A. Elçi (Ed.), *Handbook of Research on Contemporary Perspectives on Web-Based Systems* (pp. 126–141). IGI Global. doi:10.4018/978-1-5225-5384-7.ch007

Hong, J. Y., Lee, P. J., & Jeon, J. Y. (2018). Assessments of Urban Soundscapes. In F. Aletta & J. Xiao (Eds.), *Handbook of Research on Perception-Driven Approaches to Urban Assessment and Design* (pp. 46–72). IGI Global. doi:10.4018/978-1-5225-3637-6.ch003

Hossain, K., Rahman, M., & Roy, S. (2019). IoT Data Compression and Optimization Techniques in Cloud Storage. *International Journal of Cloud Applications and Computing*, *9*(2), 43–59. doi:10.4018/IJCAC.2019040103

Hovardaoglu, O., & Calisir-Hovardaoglu, S. (2019). Rural Aging, Demise of the Traditional Family and the Rise of New Family Structures in Rural Turkey. In Q. R. Yasser (Ed.), *Handbook of Research on Rural Sociology and Community Mobilization for Sustainable Growth* (pp. 1–21). IGI Global. doi:10.4018/978-1-5225-7158-2.ch001

Hudaib, A., & Albdour, L. (2019). Fog Computing to Serve the Internet of Things Applications. *International Journal of Fog Computing*, *2*(2), 44–56. doi:10.4018/IJFC.2019070103

Huderson, B., & Huderson, A. (2019). Urban STEM Education. In J. L. Wendt & D. L. Apugo (Eds.), *K-12 STEM Education in Urban Learning Environments* (pp. 1–24). IGI Global. doi:10.4018/978-1-5225-7814-7.ch001

Humayun, M., Jhanjhi, N. Z., Alamri, M. Z., & Khan, A. (2020). Smart Cities and Digital Governance. In V. Ponnusamy, K. Rafique, & N. Zaman (Eds.), *Employing Recent Technologies for Improved Digital Governance* (pp. 87–106). IGI Global. doi:10.4018/978-1-7998-1851-9.ch005

Hussain, M. M., & Beg, M. S. (2019). Using Vehicles as Fog Infrastructures for Transportation Cyber-Physical Systems (T-CPS). *International Journal of Software Science and Computational Intelligence*, *11*(1), 47–69. doi:10.4018/IJSSCI.2019010104

Ibrahim, A. H., Belay, D. G., Tiruneh, A. Z., & Kia, T. T. (2018). Social and Health Risks of Female Genital Mutilation for Medication and Braveness. *International Journal of Risk and Contingency Management*, *7*(1), 20–36. doi:10.4018/IJRCM.2018010102

Ibrahim, S. M. (2021). The Making of Creative Cities. In A. A. Galaby & A. A. Abdrabo (Eds.), *Handbook of Research on Creative Cities and Advanced Models for Knowledge-Based Urban Development* (pp. 173–196). IGI Global. doi:10.4018/978-1-7998-4948-3.ch009

Iliadis, L., Anezakis, V., Demertzis, K., & Mallinis, G. (2017). Hybrid Unsupervised Modeling of Air Pollution Impact to Cardiovascular and Respiratory Diseases. *International Journal of Information Systems for Crisis Response and Management*, *9*(3), 13–35. doi:10.4018/IJISCRAM.2017070102

Jácomo, L. (2018). Advertising-Cities Face to Smart-Cities. *International Journal of E-Planning Research*, *7*(3), 22–35. doi:10.4018/IJEPR.2018070102

Jagannathan, J., & Anitha Elavarasi, S. (2019). Current Trends. In D. J. Mala (Ed.), *Integrating the Internet of Things Into Software Engineering Practices* (pp. 181–198). IGI Global. doi:10.4018/978-1-5225-7790-4.ch009

Janet, B., & Raj, P. (2019). Smart City Applications. In P. Raj & S. Koteeswaran (Eds.), *Novel Practices and Trends in Grid and Cloud Computing* (pp. 274–305). IGI Global. doi:10.4018/978-1-5225-9023-1.ch016

Jaoua, A., Ben Ammar, M., & Awasthi, A. (2019). A Decision Support System for On-Demand Goods Delivery Using Shared Autonomous Electric Vehicles. *International Journal of Decision Support System Technology*, *11*(2), 72–88. doi:10.4018/IJDSST.2019040105

Jayashree, K., Abirami, R., & Babu, R. (2019). A Collaborative Approach of IoT, Big Data, and Smart City. In N. Dey & S. Tamane (Eds.), *Big Data Analytics for Smart and Connected Cities* (pp. 25–37). IGI Global. doi:10.4018/978-1-5225-6207-8.ch002

Jiang, H., Geertman, S., & Witte, P. (2020). A Sociotechnical Framework for Smart Urban Governance. *International Journal of E-Planning Research, 9*(1), 1–19. doi:10.4018/IJEPR.2020010101

Jiménez, R., Rodríguez, P. L., & Fernández, R. R. (2019). Green Spaces of the Metropolitan Area of Guadalajara. In J. G. Vargas-Hernández & J. A. Zdunek-Wielgołaska (Eds.), *Bioeconomical Solutions and Investments in Sustainable City Development* (pp. 151–179). IGI Global. doi:10.4018/978-1-5225-7958-8.ch007

Kaczorowska, A., & Pont, M. B. (2019). Modelling Urban Environments to Promote Ecosystem Services and Biodiversity. *International Journal of E-Planning Research, 8*(3), 1–12. doi:10.4018/IJEPR.2019070101

Kadam, V. G., Tamane, S. C., & Solanki, V. K. (2019). Smart and Connected Cities Through Technologies. In N. Dey & S. Tamane (Eds.), *Big Data Analytics for Smart and Connected Cities* (pp. 1–24). IGI Global. doi:10.4018/978-1-5225-6207-8.ch001

Kalloniatis, C., Kavroudakis, D., Polidoropoulou, A., & Gritzalis, S. (2019). Designing Privacy-Aware Intelligent Transport Systems. *International Journal of Applied Geospatial Research, 10*(1), 73–91. doi:10.4018/IJAGR.2019010104

Kamete, A. Y. (2018). A Concept 'Vandalised'. *International Journal of E-Planning Research, 7*(1), 1–14. doi:10.4018/IJEPR.2018010101

Kandpal, V. (2018). Shaping India's Future by Building Smart Future Sustainable Cities. *International Journal of Electronic Government Research, 14*(4), 27–38. doi:10.4018/IJEGR.2018100103

Kebede, M. F., Khan, B., Singh, N., & Singh, P. (2018). Energy Production in Smart Cities by Utilization of Kinetic Energy of Vehicles Over Speed Breaker. *International Journal of Civic Engagement and Social Change, 5*(2), 1–35. doi:10.4018/IJCESC.2018040101

Kedia, D., & Kaur, G. (2018). Examining Different Applications of Cloud-Based IoT. In P. Tomar & G. Kaur (Eds.), *Examining Cloud Computing Technologies Through the Internet of Things* (pp. 125–146). IGI Global. doi:10.4018/978-1-5225-3445-7.ch007

Kesavan, R., & Pushpa Jaculine, J. (2019). Smart IoT Meters for Smart Living. In K. Saravanan, G. Julie, & H. Robinson (Eds.), *Handbook of Research on Implementation and Deployment of IoT Projects in Smart Cities* (pp. 253–264). IGI Global. doi:10.4018/978-1-5225-9199-3.ch015

Khadse, V. M., Mahalle, P. N., & Shinde, G. R. (2020). Statistical Study of Machine Learning Algorithms Using Parametric and Non-Parametric Tests. *International Journal of Ambient Computing and Intelligence*, *11*(3), 80–105. doi:10.4018/IJACI.2020070105

Khan, A., Jhanjhi, N. Z., Humayun, M., & Ahmad, M. (2020). The Role of IoT in Digital Governance. In V. Ponnusamy, K. Rafique, & N. Zaman (Eds.), *Employing Recent Technologies for Improved Digital Governance* (pp. 128–150). IGI Global. doi:10.4018/978-1-7998-1851-9.ch007

Kılınç, A. (2020). An Analysis on Turkey's Urbanization Process in the Context of Urban Development Paradigms. In I. Akansel (Ed.), *Examining the Relationship Between Economics and Philosophy* (pp. 160–180). IGI Global. doi:10.4018/978-1-7998-1037-7.ch008

Kiryakova-Dineva, T., Kyurova, V., & Chankova, Y. (2021). Mega Events and Their Effect on Mesoeconomics. In P. C. Kostis (Ed.), *Bridging Microeconomics and Macroeconomics and the Effects on Economic Development and Growth* (pp. 208–230). IGI Global. doi:10.4018/978-1-7998-4933-9.ch011

Kone, S. M. (2019). A Planning Model for Cognitive Cities. In A. Voghera & L. La Riccia (Eds.), *Spatial Planning in the Big Data Revolution* (pp. 287–303). IGI Global. doi:10.4018/978-1-5225-7927-4.ch013

Kone, S. M. (2020). Urban Topology of Car-Free Cities. In R. M. Doheim, A. A. Farag, & E. Kamel (Eds.), *Humanizing Cities Through Car-Free City Development and Transformation* (pp. 185–203). IGI Global. doi:10.4018/978-1-7998-3507-3.ch008

Koo, H., Chun, Y., & Griffith, D. A. (2018). Modeling Positional Uncertainty Acquired Through Street Geocoding. *International Journal of Applied Geospatial Research*, *9*(4), 1–22. doi:10.4018/IJAGR.2018100101

Kortum, P., & Acemyan, C. Z. (2019). Usability Assessments of Mobile Applications as a Function of Geographic Location. *International Journal of Mobile Human Computer Interaction, 11*(1), 1–15. doi:10.4018/IJMHCI.2019010101

Kotb, K. S. (2021). The Features and Roles of Universities in Creative Cities. In A. A. Galaby & A. A. Abdrabo (Eds.), *Handbook of Research on Creative Cities and Advanced Models for Knowledge-Based Urban Development* (pp. 284–304). IGI Global. doi:10.4018/978-1-7998-4948-3.ch014

Koynova, T., Koleva, V., Dragoeva, A. P., & Natchev, N. (2019). Peri-Urban National Parks as Green Spaces for Recreation. *International Journal of Social Ecology and Sustainable Development, 10*(1), 46–58. doi:10.4018/IJSESD.2019010104

Kumar, K. D., Venkata Rathnam, T., Venkata Ramana, R., Sudhakara, M., & Poluru, R. K. (2020). Towards the Integration of Blockchain and IoT for Security Challenges in IoT. In D. S. Rajput, R. S. Thakur, & S. M. Basha (Eds.), *Transforming Businesses With Bitcoin Mining and Blockchain Applications* (pp. 45–67). IGI Global. doi:10.4018/978-1-7998-0186-3.ch003

Kumar, N., Agrawal, A., & Khan, R. A. (2018). A Novel Drainage System Using Cellular Automata to Avoid Urban Flood. *International Journal of Applied Evolutionary Computation, 9*(2), 38–51. doi:10.4018/IJAEC.2018040104

Kumar, P., Chouhan, L., & Songara, A. (2019). Cloud- and IoT-Powered Smart Connected Cities. In N. Dey & S. Tamane (Eds.), *Big Data Analytics for Smart and Connected Cities* (pp. 71–102). IGI Global. doi:10.4018/978-1-5225-6207-8.ch004

Kumar, S., & Raw, R. S. (2018). Improvement of Railway Transportation System Using IoT Applications and Services. In B. K. Mishra & R. Kumar (Eds.), *Big Data Management and the Internet of Things for Improved Health Systems* (pp. 120–141). IGI Global. doi:10.4018/978-1-5225-5222-2.ch008

Kumlu, K. B., & Tüdeş, Ş. (2018). Multi Criteria Decision Making Techniques in Urban Planning and Geology. In N. Ceryan (Ed.), *Handbook of Research on Trends and Digital Advances in Engineering Geology* (pp. 530–568). IGI Global. doi:10.4018/978-1-5225-2709-1.ch015

Kurebwa, J. (2019). Violence, Emotionally Abusive and Controlling Behaviour in Intimate Partner Relationships. *International Journal of Political Activism and Engagement, 6*(1), 39–51. doi:10.4018/IJPAE.2019010103 l'Her, G., Servières, M., & Siret, D. (2019). Citizen as Sensors' Commitment in Urban Public Action. *International Journal of E-Planning Research, 8*(4), 42–59. doi:10.4018/IJEPR.2019100103

Lakehal, A., Alti, A., Laborie, S., & Roose, P. (2020). A Semantic Agile Approach for Reconfigurable Distributed Applications in Pervasive Environments. *International Journal of Ambient Computing and Intelligence, 11*(2), 48–67. doi:10.4018/IJACI.2020040103

Laurini, R. (2019). Towards Knowledge-Based Spatial Planning. In A. Voghera & L. La Riccia (Eds.), *Spatial Planning in the Big Data Revolution* (pp. 1–15). IGI Global. doi:10.4018/978-1-5225-7927-4.ch001

Lavia, L., Witchel, H. J., Aletta, F., Steffens, J., Fiebig, A., Kang, J., Howes, C., & Healey, P. G. (2018). Non-Participant Observation Methods for Soundscape Design and Urban Planning. In F. Aletta & J. Xiao (Eds.), *Handbook of Research on Perception-Driven Approaches to Urban Assessment and Design* (pp. 73–99). IGI Global. doi:10.4018/978-1-5225-3637-6.ch004

Layeb, S., & Salem, M. B. (2019). Q Sensor for the City on the Threshold of Stress. In H. Abusaada, C. Vellguth, & A. Elshater (Eds.), *Handbook of Research on Digital Research Methods and Architectural Tools in Urban Planning and Design* (pp. 121–137). IGI Global. doi:10.4018/978-1-5225-9238-9.ch007

Lazarević, E. V., Ivanovic, M., Ristić, D., Marjanović, M., & Miličević, M. (2020). Small Urban Centers as Drivers of Daily Migrations and Agents of Transformation of Rural Hinterland. In A. Krstić-Furundžić & A. Djukić (Eds.), *Handbook of Research on Urban-Rural Synergy Development Through Housing, Landscape, and Tourism* (pp. 71–93). IGI Global. doi:10.4018/978-1-5225-9932-6.ch004

Lee, C., & Wong, K. D. (2018). Deriving a Gamified Learning-Design Framework Towards Sustainable Community Engagement and Mashable Innovations in Smart Cities. *International Journal of Knowledge and Systems Science, 9*(1), 1–22. doi:10.4018/IJKSS.2018010101

Leśniewska-Napierała, K., Napierała, T., Birdir, S. S., & Birdir, K. (2020). Smart Tourism Planning. In E. Çeltek (Ed.), *Handbook of Research on Smart Technology Applications in the Tourism Industry* (pp. 473–487). IGI Global. doi:10.4018/978-1-7998-1989-9.ch022

Lindley, T. (2018). Active Learning and the "Teaching" of Migration in Geography. In S. Dikli, B. Etheridge, & R. Rawls (Eds.), *Curriculum Internationalization and the Future of Education* (pp. 36–54). IGI Global. doi:10.4018/978-1-5225-2791-6.ch003

Lohachab, A. (2020). Bootstrapping Urban Planning. In R. C. Joshi & B. B. Gupta (Eds.), *Security, Privacy, and Forensics Issues in Big Data* (pp. 217–246). IGI Global. doi:10.4018/978-1-5225-9742-1.ch009

Lokshina, I. V., Durkin, B. J., & Lanting, C. (2018). The IoT- and Big Data-Driven Data Analysis Services. *International Journal of Knowledge Management*, *14*(4), 88–107. doi:10.4018/IJKM.2018100106

Lokshina, I. V., & Lanting, C. J. (2018). Qualitative Evaluation of IoT-Driven eHealth. *International Journal of Interdisciplinary Telecommunications and Networking*, *10*(4), 26–45. doi:10.4018/IJITN.2018100102

Lokshina, I. V., Lanting, C. J., & Durkin, B. J. (2018). IoT- and Big Data-Driven Data Analysis Services for Third Parties, Strategic Implications and Business Opportunities. *International Journal of Social Ecology and Sustainable Development*, *9*(3), 34–52. doi:10.4018/IJSESD.2018070103

López Millán, V. M. (2020). Software-Defined Networking. In G. Cornetta, A. Touhafi, & G. Muntean (Eds.), *Social, Legal, and Ethical Implications of IoT, Cloud, and Edge Computing Technologies* (pp. 1–27). IGI Global. doi:10.4018/978-1-7998-3817-3.ch001

Lousada, S. A., & Castanho, R. A. (2020). Cooperation Strategies Towards Sustainability in Insular Territories. In R. A. Castanho (Ed.), *Cross-Border Cooperation (CBC) Strategies for Sustainable Development* (pp. 216–238). IGI Global. doi:10.4018/978-1-7998-2513-5.ch012

Lovato, A. (2018). The Transmedia Script for Nonfictional Narratives. In R. R. Gambarato & G. C. Alzamora (Eds.), *Exploring Transmedia Journalism in the Digital Age* (pp. 235–252). IGI Global. doi:10.4018/978-1-5225-3781-6.ch014

Lu, J., Zhao, J., & Jiang, H. (2017). Spatial Distribution Characteristics and Influencing Factors of Urban Residents' Travel Carbon Emissions in Guangzhou. *International Journal of Applied Logistics*, *7*(2), 41–51. doi:10.4018/IJAL.2017070103

Luk, S. C. (2018). Universal Health Insurance Reform in China. *International Journal of Public and Private Perspectives on Healthcare, Culture, and the Environment*, *2*(2), 1–24. doi:10.4018/IJPPPHCE.2018070101

Luo, W., Li, H., & Zou, H. (2020). How to Integrate Universities and Cities Through Local Spatial Developments. *International Journal of E-Planning Research*, *9*(1), 63–86. doi:10.4018/IJEPR.2020010104

Luter, G., & Taylor, H. L. (2020). Building the Neighborly Community in the Age of Trump. In R. F. Kronick (Ed.), *Emerging Perspectives on Community Schools and the Engaged University* (pp. 98–113). IGI Global. doi:10.4018/978-1-7998-0280-8.ch006

Lwin, K. K., & Sekimoto, Y. (2018). Mapping the Spatial Distribution Patterns of Personal Time Spent Based on Trip Purpose. *International Journal of Applied Geospatial Research*, *9*(2), 1–13. doi:10.4018/IJAGR.2018040101

Lyu, K. (2021). The Development of Smart Cities in the World. In K. Lyu, M. Hu, J. Du, & V. Sugumaran (Eds.), *AI-Based Services for Smart Cities and Urban Infrastructure* (pp. 23–46). IGI Global. doi:10.4018/978-1-7998-5024-3.ch002

Lyu, K., & Hao, M. (2021). Definition and History of Smart Cities. In K. Lyu, M. Hu, J. Du, & V. Sugumaran (Eds.), *AI-Based Services for Smart Cities and Urban Infrastructure* (pp. 1–22). IGI Global. doi:10.4018/978-1-7998-5024-3.ch001

Lyu, K., & Hao, M. (2021). Policy Recommendations on the Application of AI to the Development of Smart Cities. In K. Lyu, M. Hu, J. Du, & V. Sugumaran (Eds.), *AI-Based Services for Smart Cities and Urban Infrastructure* (pp. 215–227). IGI Global. doi:10.4018/978-1-7998-5024-3.ch010

Ma, L. (2018). Digital Divide and Citizen Use of E-Government in China's Municipalities. *International Journal of Public Administration in the Digital Age*, *5*(3), 16–31. doi:10.4018/IJPADA.2018070102

Machado, R., & Azevedo, A. (2020). Determinants and Consequences of Citizens' E-Participation. *International Journal of E-Planning Research*, *9*(1), 20–43. doi:10.4018/IJEPR.2020010102

Madakam, S., & Holmukhe, R. M. (2019). Songdo Smart City. In N. Dey & S. Tamane (Eds.), *Big Data Analytics for Smart and Connected Cities* (pp. 278–298). IGI Global. doi:10.4018/978-1-5225-6207-8.ch012

Maestre-Gongora, G. P., & Bernal, W. N. (2019). Conceptual Model of Information Technology Management for Smart Cities. *Journal of Global Information Management*, *27*(2), 159–175. doi:10.4018/JGIM.2019040109

Magableh, A. A., Almakhadmeh, M. A., Alsrehin, N., & Klaib, A. F. (2020). Smart Traffic Light Management Systems. *International Journal of Technology Diffusion*, *11*(3), 22–47. doi:10.4018/IJTD.2020070102

Magrane, E. (2019). Applying the Geohumanities. *International Journal of Applied Geospatial Research*, *10*(2), 27–38. doi:10.4018/IJAGR.2019040103

Majumder, P., & Saha, A. K. (2019). Identification of Most Significant Parameter of Impact of Climate Change and Urbanization on Operational Efficiency of Hydropower Plant. *International Journal of Energy Optimization and Engineering*, *8*(3), 43–68. doi:10.4018/IJEOE.2019070103

Majumder, P., & Saha, A. K. (2019). Ranking of Indicators for Estimation of Plant Efficiency in Hydropower Plants by a Bootstrap MCDM Approach. *International Journal of Energy Optimization and Engineering*, *8*(3), 69–92. doi:10.4018/IJEOE.2019070104

Makwerere, D. (2019). African Youth's Democratic Rights Awareness and Participation in Continental Governance. In J. Kurebwa & O. Dodo (Eds.), *Participation of Young People in Governance Processes in Africa* (pp. 176–197). IGI Global. doi:10.4018/978-1-5225-9388-1.ch009

Marchigiani, E. (2020). An Accessible City is a Healthy and People-Centred Smart City. *International Journal of Urban Planning and Smart Cities*, *1*(2), 59–79. doi:10.4018/IJUPSC.2020070105

Marconi, A., & Loria, E. (2020). Sustainable Mobility in Smart Cities. In Implications of Mobility as a Service (MaaS) in Urban and Rural Environments (pp. 211-246). IGI Global. doi:10.4018/978-1-7998-1614-0.ch008

Martinelli, N., & Mangialardi, G. (2018). Cities With Grid Layout. In G. Carlone, N. Martinelli, & F. Rotondo (Eds.), *Designing Grid Cities for Optimized Urban Development and Planning* (pp. 188–208). IGI Global. doi:10.4018/978-1-5225-3613-0.ch011

Martucci, O., Acampora, A., Arcese, G., & Poponi, S. (2020). The Development of Smart Tourism Destinations Through the Integration of ICT Innovations in SMEs of the Commercial Sector. In J. M. Palma-Ruiz, J. Saiz-Álvarez, & Á. Herrero-Crespo (Eds.), *Handbook of Research on Smart Territories and Entrepreneurial Ecosystems for Social Innovation and Sustainable Growth* (pp. 124–151). IGI Global. doi:10.4018/978-1-7998-2097-0.ch008

Matthews, L. (2020). Anamorphic Atmospheres. *International Journal of Creative Interfaces and Computer Graphics, 11*(2), 27–41. doi:10.4018/IJCICG.2020070103

McMeel, D. J. (2018). The Dark Web of Urban Data. *International Journal of Art, Culture and Design Technologies, 7*(2), 12–25. doi:10.4018/IJACDT.2018070102

McNutt, J. G., & Goldkind, L. (2020). Civic Technology and Data for Good. In Digital Government and Achieving E-Public Participation (pp. 124-142). IGI Global. doi:10.4018/978-1-7998-1526-6.ch006

Medhekar, A., & Haq, F. (2018). Urbanization and New Jobs Creation in Healthcare Services in India. In U. Benna & I. Benna (Eds.), *Urbanization and Its Impact on Socio-Economic Growth in Developing Regions* (pp. 198–218). IGI Global. doi:10.4018/978-1-5225-2659-9.ch010

Megdadi, K., Akkaya, M., & Sari, A. (2018). Internet of Things and Smart City Initiatives in Middle Eastern Countries. In P. Raj & A. Raman (Eds.), *Handbook of Research on Cloud and Fog Computing Infrastructures for Data Science* (pp. 289–311). IGI Global. doi:10.4018/978-1-5225-5972-6.ch014

Melo, L. M., Carvalho, J. C., & Sabino, E. M. (2019). Orientation for Results. In J. Carvalho & E. Sabino (Eds.), *Strategy and Superior Performance of Micro and Small Businesses in Volatile Economies* (pp. 281–302). IGI Global. doi:10.4018/978-1-5225-7888-8.ch017

Mendez, C., Lazarte, M., & Cacho, R. (2019). Localization Patterns of Urban Food Distribution Centers in Metropolitan Lima. In J. Gonzalez-Feliu, M. Chong, J. Vargas Florez, & J. Padilla Solis (Eds.), *Handbook of Research on Urban and Humanitarian Logistics* (pp. 38–67). IGI Global. doi:10.4018/978-1-5225-8160-4.ch003

Mendoza, L. C., Cruz, G. A., Ciencia, A. N., & Penalba, M. A. (2020). Local Policy and Water Access in Baguio City, Philippines. *International Journal of Social Ecology and Sustainable Development, 11*(1), 1–13. doi:10.4018/IJSESD.2020010101

Meshur, H. F. (2018). Understanding Smart City Solutions in Turkish Cities From the Perspective of Sustainability. In Z. H. Gontar (Ed.), *Smart Grid Analytics for Sustainability and Urbanization* (pp. 236–266). IGI Global. doi:10.4018/978-1-5225-3996-4.ch009

Messina, J., Suepa, T., Snapp, S., Olson, J., Nejadhashemi, A. P., Murray, S., Moore, N., Frake, A., Fan, P., & Adhikari, U. (2017). Food System Resilience and Sustainability in Cambodia. *International Journal of Applied Geospatial Research, 8*(3), 53–75. doi:10.4018/ijagr.2017070104

Mikaeili, M., & Aytuğ, H. K. (2020). Analysing of Erzurum Winter Tourism in the European Perspective. In I. Akansel (Ed.), *Comparative Approaches to Old and New Institutional Economics* (pp. 88–107). IGI Global. doi:10.4018/978-1-7998-0333-1.ch006

Milo, A., Bryan-Kinns, N., & Reiss, J. D. (2018). Graphical Research Tools for Acoustic Design Training. In F. Aletta & J. Xiao (Eds.), *Handbook of Research on Perception-Driven Approaches to Urban Assessment and Design* (pp. 397–434). IGI Global. doi:10.4018/978-1-5225-3637-6.ch017

Milutinovic, G., Ahonen-Jonnarth, U., & Seipel, S. (2018). GISwaps. *International Journal of Decision Support System Technology, 10*(3), 57–78. doi:10.4018/IJDSST.2018070104

Minyoi, M. K. (2019). The Historiography of Town Planning and Its Shifting Paradigms in Sub-Saharan Africa. In I. Biao (Ed.), *Learning Cities, Town Planning, and the Creation of Livelihoods* (pp. 1–32). IGI Global. doi:10.4018/978-1-5225-8134-5.ch001

Mishra, S., Tripathy, H. K., Mishra, B. K., & Mohapatra, S. K. (2018). A Succinct Analysis of Applications and Services Provided by IoT. In B. K. Mishra & R. Kumar (Eds.), *Big Data Management and the Internet of Things for Improved Health Systems* (pp. 142–162). IGI Global. doi:10.4018/978-1-5225-5222-2.ch009

Miśkowiec, M., & Gorczyca, K. M. (2018). Public Participation in Local Regeneration Programmes in Poland. *International Journal of E-Planning Research, 7*(4), 44–60. doi:10.4018/IJEPR.2018100103

Mkrttchian, V., Vertakova, Y., Treshchevsky, Y., Firsova, N. V., Plotnikov, V., & Treshchevsky, D. Y. (2019). Smart City. In U. G. Benna (Ed.), *Industrial and Urban Growth Policies at the Sub-National, National, and Global Levels* (pp. 300–319). IGI Global. doi:10.4018/978-1-5225-7625-9.ch015

Mohamed, A. A. (2019). Space Syntax Approach for Articulating Space and Social Life. In H. Abusaada, C. Vellguth, & A. Elshater (Eds.), *Handbook of Research on Digital Research Methods and Architectural Tools in Urban Planning and Design* (pp. 223–249). IGI Global. doi:10.4018/978-1-5225-9238-9.ch011

Mohamed, K. Y., Mostafa, A. M., & el-Hameed, J. A. (2018). The Comparative Importance of Urban and Architectural Potentials Supporting Knowledge Cities Formation and Their Relevant Indices. In A. A. Alraouf (Ed.), *Knowledge-Based Urban Development in the Middle East* (pp. 35–58). IGI Global. doi:10.4018/978-1-5225-3734-2.ch003

Mohammadi, A. (2019). Determining Appropriate Zones for Knowledge Intensive Firms Site Selection Using GIS. *International Journal of Applied Geospatial Research, 10*(1), 1–30. doi:10.4018/IJAGR.2019010101

Molebatsi, C., & Morobolo, S. (2019). Debating the Informal Sector and Urban Planning in Botswana. In I. Biao (Ed.), *Learning Cities, Town Planning, and the Creation of Livelihoods* (pp. 33–54). IGI Global. doi:10.4018/978-1-5225-8134-5.ch002

Mondschein, A., Zhang, Z., & El Khafif, M. (2019). Community-Centered Urban Sensing. *International Journal of E-Planning Research, 8*(4), 1–16. doi:10.4018/IJEPR.2019100101

Montejo, G. A., & Frazier, A. E. (2020). Assessing Quality of Life Across Mexico City Using Socio-Economic and Environmental Factors. *International Journal of Applied Geospatial Research, 11*(3), 68–86. doi:10.4018/IJAGR.2020070105

Mora, O. B., Cedillo-Elias, E. J., Aceves, E., & Larios, V. M. (2019). Relevance of Technologies for Smart Cities. In J. J. Rodrigues, A. Gawanmeh, K. Saleem, & S. Parvin (Eds.), *Smart Devices, Applications, and Protocols for the IoT* (pp. 165–188). IGI Global. doi:10.4018/978-1-5225-7811-6.ch008

Morganti, C., & Bartolomei, C. (2019). 3D Digitization of Architectural Heritage. In C. Inglese & A. Ippolito (Eds.), *Analysis, Conservation, and Restoration of Tangible and Intangible Cultural Heritage* (pp. 51–85). IGI Global. doi:10.4018/978-1-5225-6936-7.ch003

Msuya, M. S. (2018). ICT for Improving Financial Access in Informal Sector. *International Journal of ICT Research in Africa and the Middle East*, *7*(2), 17–30. doi:10.4018/IJICTRAME.2018070102

Mueller, J., Asada, S., & Tomarchio, L. (2020). Engaging the Crowd. *International Journal of E-Planning Research*, *9*(2), 66–79. doi:10.4018/IJEPR.2020040101.oa

Mukherjee, F. (2020). GIS and E-Governance for Urban Local Bodies in India. In C. N. Silva (Ed.), *Citizen-Responsive Urban E-Planning* (pp. 168–193). IGI Global. doi:10.4018/978-1-7998-4018-3.ch007

Munger, M. H., Murray, M., Richardson, M., & Claussen, A. (2018). Transformative Learning in Teacher Education. *International Journal of Adult Vocational Education and Technology*, *9*(4), 54–64. doi:10.4018/IJAVET.2018100105

Muñoz de Luna, A. B., & Kolotouchkina, O. (2020). Smart Place Making Through Digital Communication and Citizen Engagement. In Digital Government and Achieving E-Public Participation (pp. 206-228). IGI Global. doi:10.4018/978-1-7998-1526-6.ch010

Musambo, L. K., & Phiri, J. (2019). A Framework for Civil Registration in Developing Countries Based on Biometrics and ISO Standards. *International Journal of ICT Research in Africa and the Middle East*, *8*(2), 32–74. doi:10.4018/IJICTRAME.2019070103

Musti, K. S. (2020). Circular Economy in Energizing Smart Cities. In N. Baporikar (Ed.), *Handbook of Research on Entrepreneurship Development and Opportunities in Circular Economy* (pp. 251–269). IGI Global. doi:10.4018/978-1-7998-5116-5.ch013

Nair, R., Sharma, P., Bhagat, A., & Dwivedi, V. K. (2018). A Survey on IoT (Internet of Things) Emerging Technologies and Its Application. *International Journal of End-User Computing and Development*, *7*(2), 1–20. doi:10.4018/IJEUCD.2018070101

Nam, T. (2019). The Evolution From Ubiquitous to Smart Cities. *International Journal of Electronic Government Research, 15*(3), 59–71. doi:10.4018/IJEGR.2019070104

Namiot, D., & Sneps-Sneppe, M. (2020). Internet of Things and Cyber-Physical Systems at the University. In S. Balandin & E. Balandina (Eds.), *Tools and Technologies for the Development of Cyber-Physical Systems* (pp. 285–302). IGI Global. doi:10.4018/978-1-7998-1974-5.ch011

Nanda, M., & Randhawa, G. (2019). Smart Cities Project. In K. Saravanan, G. Julie, & H. Robinson (Eds.), *Handbook of Research on Implementation and Deployment of IoT Projects in Smart Cities* (pp. 80–95). IGI Global. doi:10.4018/978-1-5225-9199-3.ch006

Nangia, M., & Saldanha, S. R. (2020). Role of Entrepreneurs in Promoting Sustainability in India. In G. Jain, H. Singh, S. Akter, A. Munjal, & H. S. Grewal (Eds.), *Technological Innovations for Sustainability and Business Growth* (pp. 30–38). IGI Global. doi:10.4018/978-1-5225-9940-1.ch003

Narayanan, L. K., Sankaranarayanan, S., Rodrigues, J. J., & Lorenz, P. (2020). Multi-Agent-Based Modeling for Underground Pipe Health and Water Quality Monitoring for Supplying Quality Water. *International Journal of Intelligent Information Technologies, 16*(3), 52–79. doi:10.4018/IJIIT.2020070103

Naugle, A. B., Backus, G. A., Tidwell, V. C., Kistin-Keller, E., & Villa, D. L. (2019). A Regional Model of Climate Change and Human Migration. *International Journal of System Dynamics Applications, 8*(1), 1–22. doi:10.4018/IJSDA.2019010101

Nayyar, A., Jain, R., Mahapatra, B., & Singh, A. (2019). Cyber Security Challenges for Smart Cities. In K. Ahuja & A. Khosla (Eds.), *Driving the Development, Management, and Sustainability of Cognitive Cities* (pp. 27–54). IGI Global. doi:10.4018/978-1-5225-8085-0.ch002

Ngossaha, J. M., Ngouna, R. H., Archimède, B., Patrascu, R. G., Petrisor, A., & Ndjodo, M. F. (2020). Methodological Framework for Defining the Sustainability Management Process for Urban Mobility Systems Based on System Engineering. *International Journal of Digital Innovation in the Built Environment, 9*(1), 1–21. doi:10.4018/IJDIBE.2020010101

Nguyen, C. T., & Hoang, D. B. (2020). Large-Scale Software-Defined IoT Platform for Provisioning IoT Services on Demand. *International Journal of Smart Sensor Technologies and Applications*, *1*(1), 42–64. doi:10.4018/ijssta.20200101.oa1

Ngwenya, B., Oosthuizen, J., Cross, M., Frimpong, K., & Chaibva, C. N. (2018). A Review of Heat Stress Policies in the Context of Climate Change and Its Impacts on Outdoor Workers. *International Journal of Social Ecology and Sustainable Development*, *9*(1), 1–11. doi:10.4018/IJSESD.2018010101

Nikezić, A., Ristić Trajković, J., & Milovanović, A. (2020). Urban-Rural Synergy Between Housing Spatial Patterns and Landscape. In A. Krstić-Furundžić & A. Djukić (Eds.), *Handbook of Research on Urban-Rural Synergy Development Through Housing, Landscape, and Tourism* (pp. 115–137). IGI Global. doi:10.4018/978-1-5225-9932-6.ch006

Nóbrega, R., Jacob, J., Coelho, A., Ribeiro, J., Weber, J., & Ferreira, S. (2018). Leveraging Pervasive Games for Tourism. *International Journal of Creative Interfaces and Computer Graphics*, *9*(1), 1–14. doi:10.4018/IJCICG.2018010101

Ntseane, P. G., & Biao, I. (2019). Learning Cities. In I. Biao (Ed.), *Learning Cities, Town Planning, and the Creation of Livelihoods* (pp. 73–93). IGI Global. doi:10.4018/978-1-5225-8134-5.ch004

Nummi, P., Eräranta, S., & Kahila-Tani, M. (2018). Enhancing E-Participation in Urban Planning Competitions. In C. Nunes Silva (Ed.), *New Approaches, Methods, and Tools in Urban E-Planning* (pp. 60–94). IGI Global. doi:10.4018/978-1-5225-5999-3.ch003

Oberman, T., Šćitaroci, B. B., & Jambrošić, K. (2018). Towards a Virtual Soundwalk. In F. Aletta & J. Xiao (Eds.), *Handbook of Research on Perception-Driven Approaches to Urban Assessment and Design* (pp. 317–343). IGI Global. doi:10.4018/978-1-5225-3637-6.ch014

Ohara, S. U., Milton, A. D., & Jeffery, T. D. (2018). Local Food and Fitness. In Changing Urban Landscapes Through Public Higher Education (pp. 144-169). IGI Global. doi:10.4018/978-1-5225-3454-9.ch007

Okay, E. (2018). Investment on Heat Pumps. In U. Akkucuk (Ed.), *Handbook of Research on Supply Chain Management for Sustainable Development* (pp. 194–217). IGI Global. doi:10.4018/978-1-5225-5757-9.ch011

Olalla-Caballero, B. (2020). Quality in E-Environment Development and Sustainability of Smart Cities. In J. M. Palma-Ruiz, J. Saiz-Álvarez, & Á. Herrero-Crespo (Eds.), *Handbook of Research on Smart Territories and Entrepreneurial Ecosystems for Social Innovation and Sustainable Growth* (pp. 37–52). IGI Global. doi:10.4018/978-1-7998-2097-0.ch003

Onuoha, M. E. (2019). Urbanization and Its Implication for Sustainable Development in a Circular Economy. In R. A. Turvey & S. Kurissery (Eds.), *Intellectual, Scientific, and Educational Influences on Sustainability Research* (pp. 138–167). IGI Global. doi:10.4018/978-1-5225-7302-9.ch007

Opromolla, A., & Volpi, V. (2020). Cities as Complex Systems. *International Journal of Urban Planning and Smart Cities*, *1*(2), 1–16. doi:10.4018/IJUPSC.2020070101

Ottenburger, S. S., Münzberg, T., & Strittmatter, M. (2017). Smart Grid Topologies Paving the Way for an Urban Resilient Continuity Management. *International Journal of Information Systems for Crisis Response and Management*, *9*(4), 1–22. doi:10.4018/IJISCRAM.2017100101

Özen, I. A. (2020). Internet of Things in Tourism. In E. Çeltek (Ed.), *Handbook of Research on Smart Technology Applications in the Tourism Industry* (pp. 131–154). IGI Global. doi:10.4018/978-1-7998-1989-9.ch007

Pal, A. K., & Sengupta, A. (2019). Urbanization and Ownership of Polluting Vehicles. In R. C. Das (Ed.), *Handbook of Research on Economic and Political Implications of Green Trading and Energy Use* (pp. 340–354). IGI Global. doi:10.4018/978-1-5225-8547-3.ch019

Panagiotopoulou, M., Kokla, M., & Stratigea, A. (2019). Conceptualizing Small and Medium-Sized Smart Cities in the Mediterranean Region. *International Journal of E-Planning Research*, *8*(4), 17–41. doi:10.4018/IJEPR.2019100102

Panagiotopoulou, M., Stratigea, A., & Leka, A. (2020). Gathering Global Intelligence for Assessing Performance of Smart, Sustainable, Resilient, and Inclusive Cities (S2RIC). In C. N. Silva (Ed.), *Citizen-Responsive Urban E-Planning* (pp. 305–345). IGI Global. doi:10.4018/978-1-7998-4018-3.ch012

Panchatcharam, P., & Vivekanandan, S. (2019). Internet of Things (IOT) in Healthcare – Smart Health and Surveillance, Architectures, Security Analysis and Data Transfer. *International Journal of Software Innovation*, *7*(2), 21–40. doi:10.4018/IJSI.2019040103

Pandey, K., Narula, A., Pandey, D., & Raw, R. S. (2021). An Approach Towards Intelligent Traffic Environment Using Machine Learning Algorithms. In R. S. Rao, N. Singh, O. Kaiwartya, & S. Das (Eds.), *Cloud-Based Big Data Analytics in Vehicular Ad-Hoc Networks* (pp. 1–22). IGI Global. doi:10.4018/978-1-7998-2764-1.ch001

Papadima, A., & Kourdis, E. (2018). Semiotic Landscape in Cyprus. *International Journal of Semiotics and Visual Rhetoric*, *2*(2), 1–22. doi:10.4018/IJSVR.2018070101

Papadopoulos, H. (2018). Technological Advancements and Distortion of Notion of Place. In Optimizing Current Practices in E-Services and Mobile Applications (pp. 62-83). IGI Global. doi:10.4018/978-1-5225-5026-6.ch004

Papadopoulou, C., & Hatzichristos, T. (2020). Allocation of Residential Areas in Smart Insular Communities. *International Journal of E-Planning Research*, *9*(4), 40–60. doi:10.4018/IJEPR.2020100103

Papadopoulou, P., Kolomvatsos, K., & Hadjiefthymiades, S. (2020). Internet of Things in E-Government. *International Journal of Artificial Intelligence and Machine Learning*, *10*(2), 99–118. doi:10.4018/IJAIML.2020070106

Park, J. W., & Choi, S. (2019). Urban Smart Block. *International Journal of Embedded and Real-Time Communication Systems*, *10*(3), 21–31. doi:10.4018/IJERTCS.2019070102

Pasalar, C., Hallowell, G. D., & Lu, Y. (2020). Smart Cars, Smart Cities, and Smart Sharing. In R. M. Doheim, A. A. Farag, & E. Kamel (Eds.), *Humanizing Cities Through Car-Free City Development and Transformation* (pp. 71–97). IGI Global. doi:10.4018/978-1-7998-3507-3.ch003

Patel, H. C., & Kulkarni, G. R. (2018). MCFS Enhanced Route Selection Approach Using Multipath Carry Forwarding and Searching in VANET. *International Journal of Vehicular Telematics and Infotainment Systems*, *2*(1), 1–17. doi:10.4018/IJVTIS.2018010101

Patil, V. K., & Nagarale, P. (2019). Prediction of L10 and Leq Noise Levels Due to Vehicular Traffic in Urban Area Using ANN and Adaptive Neuro-Fuzzy Interface System (ANFIS) Approach. *International Journal of Business Data Communications and Networking*, *15*(2), 92–105. doi:10.4018/IJBDCN.2019070106

Patnaik, K. S., & Snigdh, I. (2019). Architectural Modelling of Cyber Physical Systems Using UML. *International Journal of Cyber-Physical Systems, 1*(2), 19–37. doi:10.4018/IJCPS.2019070102

Pego, A., & Matos Bernardo, M. D. (2019). The Role of Urban Living Labs in Entrepreneurship, Energy, and Governance of Smart Cities. In L. C. Carvalho & P. Isaías (Eds.), *Handbook of Research on Entrepreneurship and Marketing for Global Reach in the Digital Economy* (pp. 203–221). IGI Global. doi:10.4018/978-1-5225-6307-5.ch009

Pereira, G. V., Eibl, G., Stylianou, C., Martínez, G., Neophytou, H., & Parycek, P. (2018). The Role of Smart Technologies to Support Citizen Engagement and Decision Making. *International Journal of Electronic Government Research, 14*(4), 1–17. doi:10.4018/IJEGR.2018100101

Pérez-de-Azpillaga, L. G., Fernández-Tabales, A., & Foronda-Robles, C. (2020). Intelligent Tourist Destinations and Their Application to Public Policies. In E. Çeltek (Ed.), *Handbook of Research on Smart Technology Applications in the Tourism Industry* (pp. 447–472). IGI Global. doi:10.4018/978-1-7998-1989-9.ch021

Pettenati, G., Dansero, E., & Calafiore, A. (2019). Researching and Enabling Youth Geographies in the Digital and Material City. In A. Voghera & L. La Riccia (Eds.), *Spatial Planning in the Big Data Revolution* (pp. 221–247). IGI Global. doi:10.4018/978-1-5225-7927-4.ch010

Pinterič, U. (2020). Human Factor and ICT Use in the Context of Modern Governance. In Digital Government and Achieving E-Public Participation (pp. 20-38). IGI Global. doi:10.4018/978-1-7998-1526-6.ch002

Policek, N. (2019). "Asking the Woman Question" in Case Study Research. In Case Study Methodology in Higher Education (pp. 298-322). IGI Global. doi:10.4018/978-1-5225-9429-1.ch014

Pombo, L., Marques, M. M., Afonso, L., Dias, P., & Madeira, J. (2019). Evaluation of a Mobile Augmented Reality Game Application as an Outdoor Learning Tool. *International Journal of Mobile and Blended Learning, 11*(4), 59–79. doi:10.4018/IJMBL.2019100105

Popescu, C. R. (2020). Approaches to Sustainable and Responsible Entrepreneurship. In B. R. Hernández-Sánchez, J. C. Sánchez-García, & A. C. Moreira (Eds.), *Building an Entrepreneurial and Sustainable Society* (pp. 75–95). IGI Global. doi:10.4018/978-1-7998-2704-7.ch004

Popoola, A. A., & Magidimisha, H. H. (2020). Investigating the Roles Played by Selected Agencies in Infrastructure Development. In M. A. Mafukata & K. A. Tshikolomo (Eds.), *African Perspectives on Reshaping Rural Development* (pp. 289–319). IGI Global. doi:10.4018/978-1-7998-2306-3.ch014

Prasetyo, F. A. (2020). The Car-Free (Day) Movement. In R. M. Doheim, A. A. Farag, & E. Kamel (Eds.), *Humanizing Cities Through Car-Free City Development and Transformation* (pp. 115–138). IGI Global. doi:10.4018/978-1-7998-3507-3.ch005

Provis, E. L. (2018). Implementing Reform of Legacy Housing Stock Against Bushfire. *International Journal of Disaster Response and Emergency Management*, *1*(2), 23–38. doi:10.4018/IJDREM.2018040102

Puri, V., Van Le, C., Kumar, R., & Jagdev, S. S. (2020). Fruitful Synergy Model of Artificial Intelligence and Internet of Thing for Smart Transportation System. *International Journal of Hyperconnectivity and the Internet of Things*, *4*(1), 43–57. doi:10.4018/IJHIoT.2020010104

Purwanto, E. A. (2018). Smart City as an Upshot of Bureaucratic Reform in Indonesia. *International Journal of Electronic Government Research*, *14*(3), 32–43. doi:10.4018/IJEGR.2018070103

Putra, Z. D., & van der Knaap, W. G. (2020). Pasikola. *International Journal of E-Planning Research*, *9*(3), 24–46. doi:10.4018/IJEPR.2020070102

Qiu, X., Sha, D., & Meng, X. (2019). Optimal Methodology for Detecting Land Cover Change in a Forestry, Lakeside Environment Using NAIP Imagery. *International Journal of Applied Geospatial Research*, *10*(1), 31–53. doi:10.4018/IJAGR.2019010102

Ranjan, R., Chatterjee, P., Panchal, D., & Pamucar, D. (2019). Performance Evaluation of Sustainable Smart Cities in India. In P. Chatterjee, M. Yazdani, S. Chakraborty, D. Panchal, & S. Bhattacharyya (Eds.), *Advanced Multi-Criteria Decision Making for Addressing Complex Sustainability Issues* (pp. 14–40). IGI Global. doi:10.4018/978-1-5225-8579-4.ch002

Rashid, M. F. (2020). Spatial Modelling of Migration Decision-Selectivity in the Klang Valley Region, Malaysia. *International Journal of Applied Geospatial Research*, *11*(1), 21–35. doi:10.4018/IJAGR.2020010102

Rath, M. (2019). Smart Traffic Management and Secured Framework for Smart Cities. In R. A. Shaikh (Ed.), *Secure Cyber-Physical Systems for Smart Cities* (pp. 27–51). IGI Global. doi:10.4018/978-1-5225-7189-6.ch002

Rath, M. (2021). Big Data Analytics and IoT in Smart City Applications. In Encyclopedia of Information Science and Technology, Fifth Edition (pp. 586-601). IGI Global. doi:10.4018/978-1-7998-3479-3.ch042

Rath, M., & Misra, R. (2018). An Exhaustive Study and Analysis of Assorted Application and Challenges in Fog Computing and Emerging Ubiquitous Computing Technology. *International Journal of Applied Evolutionary Computation*, *9*(2), 17–32. doi:10.4018/IJAEC.2018040102

Rathee, D. S., Ahuja, K., & Hailu, T. (2019). Role of Electronics Devices for E-Health in Smart Cities. In K. Ahuja & A. Khosla (Eds.), *Driving the Development, Management, and Sustainability of Cognitive Cities* (pp. 212–233). IGI Global. doi:10.4018/978-1-5225-8085-0.ch009

Rathore, M. M., Ahmad, A., & Paul, A. (2018). Big Data Deep Analytics for Geosocial Networks. In S. Karthik, A. Paul, & N. Karthikeyan (Eds.), *Deep Learning Innovations and Their Convergence With Big Data* (pp. 120–140). IGI Global. doi:10.4018/978-1-5225-3015-2.ch007

Ravi, N., Vimala Rani, P., Rajesh Alias Harinarayan, R., Mercy Shalinie, S., Seshadri, K., & Pariventhan, P. (2019). Deep Learning-based Framework for Smart Sustainable Cities. *International Journal of Intelligent Information Technologies*, *15*(4), 76–107. doi:10.4018/IJIIT.2019100105

Rawat, P., & Yusuf, J. W. (2020). Participatory Mapping, E-Participation, and E-Governance. In N. Vij Mali (Ed.), *Leveraging Digital Innovation for Governance, Public Administration, and Citizen Services* (pp. 147–175). IGI Global. doi:10.4018/978-1-5225-5412-7.ch007

Recalde, J. M., Palau, R., Galés, N. L., & Gallon, R. (2020). Developments for Smart Classrooms. *International Journal of Mobile and Blended Learning*, *12*(4), 34–50. doi:10.4018/IJMBL.2020100103

Refaat, A. M. (2019). Space Syntax Theory and Its Contribution to Urban Design. In H. Abusaada, C. Vellguth, & A. Elshater (Eds.), *Handbook of Research on Digital Research Methods and Architectural Tools in Urban Planning and Design* (pp. 203–222). IGI Global. doi:10.4018/978-1-5225-9238-9.ch010

Rehena, Z., Janssen, M., & Chattopadhyay, S. (2018). A Reference Architecture for Context-Aware Intelligent Traffic Management Platforms. *International Journal of Electronic Government Research, 14*(4), 65–79. doi:10.4018/IJEGR.2018100105

Reji, A. A., & Muruganantham, S. (2019). Building Detection From Satellite Images for Urban Planning Using MATLAB-Based Pattern Matching Method. *International Journal of Creative Interfaces and Computer Graphics, 10*(2), 17–28. doi:10.4018/IJCICG.2019070102

Ren, X., & Wang, D. (2020). Research on Collective Human Mobility in Shanghai Based on Cell Phone Data. *International Journal of E-Planning Research, 9*(1), 44–62. doi:10.4018/IJEPR.2020010103

Requejo Flores, A., Ruiz, A., López, A., & Porras, R. (2021). News Classification to Notify About Traffic Incidents in a Mexican City. In R. A. Pazos-Rangel, R. Florencia-Juarez, M. A. Paredes-Valverde, & G. Rivera (Eds.), *Handbook of Research on Natural Language Processing and Smart Service Systems* (pp. 227–244). IGI Global. doi:10.4018/978-1-7998-4730-4.ch010

Requejo Flores, A., Ruiz, A., Mar, R., & Porras, R. (2021). Location Extraction to Inform a Spanish-Speaking Community About Traffic Incidents. In R. A. Pazos-Rangel, R. Florencia-Juarez, M. A. Paredes-Valverde, & G. Rivera (Eds.), *Handbook of Research on Natural Language Processing and Smart Service Systems* (pp. 347–367). IGI Global. doi:10.4018/978-1-7998-4730-4.ch016

Riascos, A. J., & Serna, N. (2018). Machine Learning Based Program to Prevent Hospitalizations and Reduce Costs in the Colombian Statutory Health Care System. *International Journal of Knowledge Discovery in Bioinformatics, 8*(2), 44–64. doi:10.4018/IJKDB.2018070103

Rotondo, F. (2018). The Grid Cities. In G. Carlone, N. Martinelli, & F. Rotondo (Eds.), *Designing Grid Cities for Optimized Urban Development and Planning* (pp. 109–127). IGI Global. doi:10.4018/978-1-5225-3613-0.ch007

Rungvithu, T., & Kerdvibulvech, C. (2019). Conversational Commerce and CryptoCurrency Research in Urban Office Employees in Thailand. *International Journal of e-Collaboration, 15*(3), 34–48. doi:10.4018/IJeC.2019070103

Sabry, A. H., Benhra, J., & Bacha, A. (2018). ASIF Approach to Solve the Green Traveling Salesman Problem. *International Journal of Operations Research and Information Systems, 9*(1), 81–95. doi:10.4018/IJORIS.2018010105

Sadeh, A., Feniser, C., & Dusa, S. I. (2020). Technology Education and Learning in Smart Cities. In Developing Technology Mediation in Learning Environments (pp. 78-95). IGI Global. doi:10.4018/978-1-7998-1591-4.ch005

Sahar, L., Williams, R., Rao, A., Alcaraz, K. I., & Portier, K. M. (2018). Using GIS Technology to Define and Assess a Rurality Scheme Suitable for Decision Support in Health and Patient Services. *International Journal of Applied Geospatial Research, 9*(3), 1–17. doi:10.4018/IJAGR.2018070101

Sahraoui, A., Makhlouf, D., & Roose, P. (2018). Smart Traffic Management System for Anticipating Unexpected Road Incidents in Intelligent Transportation Systems. *International Journal of Grid and High Performance Computing, 10*(4), 67–82. doi:10.4018/IJGHPC.2018100104

Saini, S., & Singh, J. (2020). Managing Consumer Loyalty. *International Journal of Asian Business and Information Management, 11*(1), 21–47. doi:10.4018/IJABIM.2020010102

Salizzoni, E. (2019). Ecosystem Service Evaluation for Landscape Planning Policies. In A. Voghera & L. La Riccia (Eds.), *Spatial Planning in the Big Data Revolution* (pp. 177–192). IGI Global. doi:10.4018/978-1-5225-7927-4.ch008

San Martín, H., García-de-los-Salmones, M. M., & Herrero-Crespo, Á. (2020). Citizen Perceptions and Support for Smart City Projects. In J. M. Palma-Ruiz, J. Saiz-Álvarez, & Á. Herrero-Crespo (Eds.), *Handbook of Research on Smart Territories and Entrepreneurial Ecosystems for Social Innovation and Sustainable Growth* (pp. 107–123). IGI Global. doi:10.4018/978-1-7998-2097-0.ch007

Sanchez, G. M., Alves, S., & Botteldooren, D. (2018). Urban Sound Planning. In F. Aletta & J. Xiao (Eds.), *Handbook of Research on Perception-Driven Approaches to Urban Assessment and Design* (pp. 1–22). IGI Global. doi:10.4018/978-1-5225-3637-6.ch001

Sánchez, J. R., & Sánchez, M. J. (2018). Sensitive Bodies in the Cityscape. In F. Aletta & J. Xiao (Eds.), *Handbook of Research on Perception-Driven Approaches to Urban Assessment and Design* (pp. 260–276). IGI Global. doi:10.4018/978-1-5225-3637-6.ch011

Sánchez Martínez, M. (2020). Smart Destinations as a Reconversion Strategy for Rural Areas. In G. Cornetta, A. Touhafi, & G. Muntean (Eds.), *Social, Legal, and Ethical Implications of IoT, Cloud, and Edge Computing Technologies* (pp. 203–220). IGI Global. doi:10.4018/978-1-7998-3817-3.ch009

Sandoval, V., Sarmiento, J. P., Mazariegos, E. A., & Oviedo, D. (2020). Exploring Network Analysis for Urban Planning and Disaster Risk Reduction in Informal Settlements. *International Journal of Disaster Response and Emergency Management*, *3*(1), 30–45. doi:10.4018/IJDREM.2020010103

Sankar, S., & Srinivasan, P. (2020). Enhancing the Mobility Support in Internet of Things. *International Journal of Fuzzy System Applications*, *9*(4), 1–20. doi:10.4018/IJFSA.2020100101

Sanli, T. (2021). Aestheticization Through Representation of Power in Built Environment. In M. N. Erdem, N. Kocabay-Sener, & T. Demir (Eds.), *Handbook of Research on Aestheticization of Violence, Horror, and Power* (pp. 347–368). IGI Global. doi:10.4018/978-1-7998-4655-0.ch018

Santos, B. (2020). Urban Information Systems and Digital Channels to Enhance Transparency and Participation in Urban Planning. In C. N. Silva (Ed.), *Citizen-Responsive Urban E-Planning* (pp. 83–110). IGI Global. doi:10.4018/978-1-7998-4018-3.ch004

Santos, J. R. (2020). Decoding (Sub)Urban-Rural Fragmentation Processes. In A. Krstić-Furundžić & A. Djukić (Eds.), *Handbook of Research on Urban-Rural Synergy Development Through Housing, Landscape, and Tourism* (pp. 1–29). IGI Global. doi:10.4018/978-1-5225-9932-6.ch001

Sarowar, M. G., Kamal, M. S., & Dey, N. (2019). Internet of Things and Its Impacts in Computing Intelligence. In N. Dey & S. Tamane (Eds.), *Big Data Analytics for Smart and Connected Cities* (pp. 103–136). IGI Global. doi:10.4018/978-1-5225-6207-8.ch005

Schipper, R., & Silvius, G. (2019). Opportunities for the Circular Economy in Smart Cities. *International Journal of Information Systems and Social Change*, *10*(4), 12–35. doi:10.4018/IJISSC.2019100102

Scholl, M. C. (2019). Raising Information Security Awareness in the Field of Urban and Regional Planning. *International Journal of E-Planning Research*, *8*(3), 62–86. doi:10.4018/IJEPR.2019070104

Scuderi, A., & Sturiale, L. (2019). Defining Urban Planning Strategy Through Social Media Application. In A. Voghera & L. La Riccia (Eds.), *Spatial Planning in the Big Data Revolution* (pp. 268–286). IGI Global. doi:10.4018/978-1-5225-7927-4.ch012

Sejdiu, B., Ismaili, F., & Ahmedi, L. (2020). Integration of Semantics Into Sensor Data for the IoT. *International Journal on Semantic Web and Information Systems, 16*(4), 1–25. doi:10.4018/IJSWIS.2020100101

Sengupta, D. (2020). Taxonomy on Ambient Computing. *International Journal of Ambient Computing and Intelligence, 11*(1), 1–33. doi:10.4018/IJACI.2020010101

Sergunin, A. (2019). Russian Arctic Cities' Sustainable Development Strategies. In V. Erokhin, T. Gao, & X. Zhang (Eds.), *Handbook of Research on International Collaboration, Economic Development, and Sustainability in the Arctic* (pp. 495–511). IGI Global. doi:10.4018/978-1-5225-6954-1.ch023

Sertyesilisik, B. (2019). Circular, Smart, and Connected Cities. In K. Saravanan, G. Julie, & H. Robinson (Eds.), *Handbook of Research on Implementation and Deployment of IoT Projects in Smart Cities* (pp. 19–32). IGI Global. doi:10.4018/978-1-5225-9199-3.ch002

Sertyesilisik, B. (2019). Potential Changes in the Demand and Supply Sides in the Construction Industry. In B. Christiansen, I. Sysoeva, A. Udovikina, & A. Ketova (Eds.), *Emerging Economic Models for Global Sustainability and Social Development* (pp. 187–202). IGI Global. doi:10.4018/978-1-5225-5787-6.ch010

Shaanika, I. N. (2021). Enterprise Architecture Framework for Windhoek Smart City Realisation. In T. Iyamu (Ed.), *Empowering Businesses With Collaborative Enterprise Architecture Frameworks* (pp. 110–128). IGI Global. doi:10.4018/978-1-5225-8229-8.ch005

Shahabian, A., Fadai, A., & Peruzzi, T. (2020). Future of Life-Cycle Assessment in a Smart and/or Sustainable World. In R. Das & N. Mandal (Eds.), *Interdisciplinary Approaches to Public Policy and Sustainability* (pp. 177–207). IGI Global. doi:10.4018/978-1-7998-0315-7.ch009

Shaikh, S. A., Matono, A., & Kim, K. (2020). A Distance-Window Approach for the Continuous Processing of Spatial Data Streams. *International Journal of Multimedia Data Engineering and Management, 11*(2), 16–30. doi:10.4018/IJMDEM.2020040102

Sharma, O., & S., A. (2021). Large-Scale Data Streaming in Fog Computing and Its Applications. In *Large-Scale Data Streaming, Processing, and Blockchain Security* (pp. 50-65). IGI Global. doi:10.4018/978-1-7998-3444-1.ch003

Shinde, G. R., & Olesen, H. (2018). Beacon-Based Cluster Framework for Internet of People, Things, and Services (IoPTS). *International Journal of Ambient Computing and Intelligence*, *9*(4), 15–33. doi:10.4018/IJACI.2018100102

Shukla, P. S., & Mathur, M. (2020). Conceptualizing the Role of Data Analytics and Technology in E-Governance. *International Journal of Business Analytics*, *7*(2), 1–12. doi:10.4018/IJBAN.2020040101

Siljanoska, J., Mickovski, G., & Angelova, I. (2020). Changing Territories and the "Edge Effect" of the Re-Habited Industrial Structures. In A. Krstić-Furundžić & A. Djukić (Eds.), *Handbook of Research on Urban-Rural Synergy Development Through Housing, Landscape, and Tourism* (pp. 94–114). IGI Global. doi:10.4018/978-1-5225-9932-6.ch005

Silva, F. A., & Lopes, T. A. (2020). Application of Smart Tourism to Nature-Based Destinations. In L. C. Carvalho, L. Calisto, & N. Gustavo (Eds.), *Strategic Business Models to Support Demand, Supply, and Destination Management in the Tourism and Hospitality Industry* (pp. 1–20). IGI Global. doi:10.4018/978-1-5225-9936-4.ch001

Singh, L., & Kumar, K. (2020). Tourist Experience in Smart City Destination of Jaipur (Rajasthan). In E. Çeltek (Ed.), *Handbook of Research on Smart Technology Applications in the Tourism Industry* (pp. 430–446). IGI Global. doi:10.4018/978-1-7998-1989-9.ch020

Siountri, K., Skondras, E., & Vergados, D. D. (2020). Developing Smart Buildings Using Blockchain, Internet of Things, and Building Information Modeling. *International Journal of Interdisciplinary Telecommunications and Networking*, *12*(3), 1–15. doi:10.4018/IJITN.2020070101

Smirnov, A., Ponomarev, A., Shilov, N., Kashevnik, A., & Teslya, N. (2018). Ontology-Based Human-Computer Cloud for Decision Support. *International Journal of Embedded and Real-Time Communication Systems*, *9*(1), 1–19. doi:10.4018/IJERTCS.2018010101

Smith, E. A. (2019). Student Activism at the Urban Community College. In M. T. Miller & D. V. Tolliver (Eds.), *Exploring the Technological, Societal, and Institutional Dimensions of College Student Activism* (pp. 103–113). IGI Global. doi:10.4018/978-1-5225-7274-9.ch007

Smyth, S. J., Curran, K., & McKelvey, N. (2018). Smart Cities, Smart Grids, and Smart Grid Analytics. In Z. H. Gontar (Ed.), *Smart Grid Analytics for Sustainability and Urbanization* (pp. 103–137). IGI Global. doi:10.4018/978-1-5225-3996-4.ch004

Soeiro, D. (2020). The Concept of Atmospheres. In H. Abusaada, A. M. Salama, & A. Elshater (Eds.), *Reconstructing Urban Ambiance in Smart Public Places* (pp. 23–40). IGI Global. doi:10.4018/978-1-7998-3856-2.ch002

Sofia, K., & Ilham, K. (2018). Multi-Layer Agent Based Architecture for Internet of Things Systems. *Journal of Information Technology Research, 11*(4), 32–52. doi:10.4018/JITR.2018100103

Şoimoşan, T. M., Felseghi, R. A., Răboacă, M. S., & Filote, C. (2019). Heating Systems. In A. Tantau (Ed.), *Retrofitting for Optimal Energy Performance* (pp. 283–307). IGI Global. doi:10.4018/978-1-5225-9104-7.ch012

Solanki, A., & Nayyar, A. (2019). Green Internet of Things (G-IoT). In (Ed.), Handbook of Research on Big Data and the IoT (pp. 379-405). IGI Global. doi:10.4018/978-1-5225-7432-3.ch021

Soltani, A., & Sharifi, E. (2019). Understanding and Analysing the Urban Heat Island (UHI) Effect in Micro-Scale. *International Journal of Social Ecology and Sustainable Development, 10*(2), 14–28. doi:10.4018/IJSESD.2019040102

Somarakis, G., & Stratigea, A. (2019). Guiding Informed Choices on Participation Tools in Spatial Planning. *International Journal of E-Planning Research, 8*(3), 38–61. doi:10.4018/IJEPR.2019070103

Somasundram, S. (2020). Impact of E-Commerce on the Urban Landscape. In R. C. Ho (Ed.), *Strategies and Tools for Managing Connected Consumers* (pp. 34–45). IGI Global. doi:10.4018/978-1-5225-9697-4.ch003

Staffans, A., Kahila-Tani, M., Geertman, S., Sillanpää, P., & Horelli, L. (2020). Communication-Oriented and Process-Sensitive Planning Support. *International Journal of E-Planning Research, 9*(2), 1–20. doi:10.4018/IJEPR.2020040101

Stock, W. G., Barth, J., & Gremm, J. (2019). The Era After Oil. In M. D. Lytras, L. Daniela, & A. Visvizi (Eds.), *Knowledge-Intensive Economies and Opportunities for Social, Organizational, and Technological Growth* (pp. 63–88). IGI Global. doi:10.4018/978-1-5225-7347-0.ch005

Strize, A., Jadrić, M., & Ćukušić, M. (2020). Smart City Solutions and the Potential of Using Web APIs. In U. Akkucuk (Ed.), *Handbook of Research on Creating Sustainable Value in the Global Economy* (pp. 38–54). IGI Global. doi:10.4018/978-1-7998-1196-1.ch003

Styers, D. M., Dobbs, G. R., Cerveny, L. K., & Hayes, I. T. (2018). Geovisualization of Socio-Spatial Data on Outdoor Activities and Values in the Southern Appalachians. *International Journal of Applied Geospatial Research, 9*(3), 55–80. doi:10.4018/IJAGR.2018070104

Suárez, M. (2020). Public Services and Evolution of Smart Cities. In G. Cornetta, A. Touhafi, & G. Muntean (Eds.), *Social, Legal, and Ethical Implications of IoT, Cloud, and Edge Computing Technologies* (pp. 277–298). IGI Global. doi:10.4018/978-1-7998-3817-3.ch013

Subramanian, C. M., & Cherukuri, A. K. (2019). Fuzzy Role Based Access Control Design using Fuzzy Ontology. *International Journal of Information Technology Project Management, 10*(4), 118–136. doi:10.4018/IJITPM.2019100105

Suleiman, M. A., & Benna, U. G. (2019). The Integration of Urban Growth and Industrial Policies in Africa. In U. G. Benna (Ed.), *Industrial and Urban Growth Policies at the Sub-National, National, and Global Levels* (pp. 100–123). IGI Global. doi:10.4018/978-1-5225-7625-9.ch006

Sun, X., Fan, D., Li, Q., & Fu, B. (2021). Interpretation of the Construction Standard of Smart City Standard Systems. In K. Lyu, M. Hu, J. Du, & V. Sugumaran (Eds.), *AI-Based Services for Smart Cities and Urban Infrastructure* (pp. 91–101). IGI Global. doi:10.4018/978-1-7998-5024-3.ch004

Susha, I., Pardo, T. A., Janssen, M., Adler, N., Verhulst, S. G., & Harbour, T. (2018). A Research Roadmap to Advance Data Collaboratives Practice as a Novel Research Direction. *International Journal of Electronic Government Research, 14*(3), 1–11. doi:10.4018/IJEGR.2018070101

Swinton, A. K., & Williams, L. A. (2018). Urban Education Technological Disparities. In D. Mentor (Ed.), *Computer-Mediated Learning for Workforce Development* (pp. 41–67). IGI Global. doi:10.4018/978-1-5225-4111-0.ch003

Tallapragada, V. V. (2019). Smart IoT for Smart Cities Implementation. In K. Saravanan, G. Julie, & H. Robinson (Eds.), *Handbook of Research on Implementation and Deployment of IoT Projects in Smart Cities* (pp. 43–62). IGI Global. doi:10.4018/978-1-5225-9199-3.ch004

Tamzini, K., Hafi, Y., Ben Ouannes, A., & Borhani, R. (2021). The Role of DPPs in Promoting Local Government-Citizen Collaboration and Participation. In Encyclopedia of Information Science and Technology, Fifth Edition (pp. 1578-1596). IGI Global. doi:10.4018/978-1-7998-3479-3.ch108

Tasbasi, A., Sarıca, P. Y., & Yüksel, A. H. (2020). Scale Matters. In R. A. Castanho (Ed.), *Cross-Border Cooperation (CBC) Strategies for Sustainable Development* (pp. 86–100). IGI Global. doi:10.4018/978-1-7998-2513-5.ch005

Tataroğlu, N. (2019). Regional Integration Through Locality. In T. U. Uysal & C. Aldemir (Eds.), *Multi-Level Governance in Developing Economies* (pp. 190–214). IGI Global. doi:10.4018/978-1-5225-5547-6.ch008

Temes-Cordovez, R., Serrano-Lanzarote, B., Tuset Davó, J. J., & De León-Rodríguez, R. (2020). Alternative Tourism Strategies to Enhance Local Sustainable Tourism Development. In Handbook of Research on the Impacts, Challenges, and Policy Responses to Overtourism (pp. 261-283). IGI Global. doi:10.4018/978-1-7998-2224-0.ch014

Thilagavathi, C., Rajeswari, M., Sheethal, M. S., Devassy, D., Priya, K. V., & Divya, R. (2019). Security Issues on Internet of Things in Smart Cities. In K. Saravanan, G. Julie, & H. Robinson (Eds.), *Handbook of Research on Implementation and Deployment of IoT Projects in Smart Cities* (pp. 149–164). IGI Global. doi:10.4018/978-1-5225-9199-3.ch010

Thirumal, J., Devi, U. K., & Dynisha, S. (2019). IoT for Waste Management. In K. Saravanan, G. Julie, & H. Robinson (Eds.), *Handbook of Research on Implementation and Deployment of IoT Projects in Smart Cities* (pp. 265–278). IGI Global. doi:10.4018/978-1-5225-9199-3.ch016

Throne, R. (2021). New Investigator Fidelity. In *Practice-Based and Practice-Led Research for Dissertation Development* (pp. 165-187). IGI Global. doi:10.4018/978-1-7998-6664-0.ch008

Toimah, A. M., & Afifi, S. M. (2019). Introducing a Conceptual Framework for Spatial Decision SIM. In H. Abusaada, C. Vellguth, & A. Elshater (Eds.), *Handbook of Research on Digital Research Methods and Architectural Tools in Urban Planning and Design* (pp. 138–155). IGI Global. doi:10.4018/978-1-5225-9238-9.ch008

Tomor, Z. (2020). Citizens in the Smart City. *International Journal of Public Administration in the Digital Age*, *7*(1), 1–16. doi:10.4018/IJPADA.2020010101

Tripathi, A., Kaur, S., Sankaranarayanan, S., Narayanan, L. K., & Tom, R. J. (2019). Water Demand Prediction for Housing Apartments Using Time Series Analysis. *International Journal of Intelligent Information Technologies*, *15*(4), 57–75. doi:10.4018/IJIIT.2019100104

Tsaligopoulos, A., Economou, C., & Matsinos, Y. G. (2018). Identification, Prioritization, and Assessment of Urban Quiet Areas. In F. Aletta & J. Xiao (Eds.), *Handbook of Research on Perception-Driven Approaches to Urban Assessment and Design* (pp. 150–180). IGI Global. doi:10.4018/978-1-5225-3637-6.ch007

Tu, M., Xu, S., & Xu, J. (2018). Evaluation Method of Public Transportation System Based on Fuzzy Cloud Model. *International Journal of Digital Crime and Forensics*, *10*(4), 36–51. doi:10.4018/IJDCF.2018100103

Turcea, V. C., & Mihai, M. M. (2019). The Impact of CO2 Emissions From Waste on Human Health. *International Journal of Sustainable Economies Management*, *8*(1), 17–28. doi:10.4018/IJSEM.2019010102

Turner, R. M., Kamel, E., & Ramadan, A. (2020). Confronting the Conundrum of Shared Space Street Design. In R. M. Doheim, A. A. Farag, & E. Kamel (Eds.), *Humanizing Cities Through Car-Free City Development and Transformation* (pp. 139–163). IGI Global. doi:10.4018/978-1-7998-3507-3.ch006

Turvey, R. A. (2019). Place-Making and Sustainable Community Development. In R. A. Turvey & S. Kurissery (Eds.), *Intellectual, Scientific, and Educational Influences on Sustainability Research* (pp. 253–272). IGI Global. doi:10.4018/978-1-5225-7302-9.ch011

Turvey, R. A. (2019). Planning Urban Futures With Reference to Sustainable Cities. In R. A. Turvey & S. Kurissery (Eds.), *Intellectual, Scientific, and Educational Influences on Sustainability Research* (pp. 198–225). IGI Global. doi:10.4018/978-1-5225-7302-9.ch009

Ugurlu, D., & Sertyesilisik, B. (2019). Usage of BIM in Smart Cities. *International Journal of Digital Innovation in the Built Environment*, *8*(1), 17–27. doi:10.4018/IJDIBE.2019010102

Ushakov, D., & Chich-Jen, S. (2020). Global Economy Urbanization and Urban Economies Globalization. In D. Ushakov (Ed.), *Migration and Urbanization* (pp. 200–218). IGI Global. doi:10.4018/978-1-7998-0111-5.ch011

Usman, A. B., Gutierrez, J. A., & Bichi, A. B. (2019). Secure Routing Protocols Using Trust-Based Mechanisms in the Internet of Things for Smart City Environment Challenges and Future Trends. In R. A. Shaikh (Ed.), *Secure Cyber-Physical Systems for Smart Cities* (pp. 103–129). IGI Global. doi:10.4018/978-1-5225-7189-6.ch005

Usmani, R. S., Hashem, I. A., Pillai, T. R., Saeed, A., & Abdullahi, A. M. (2020). Geographic Information System and Big Spatial Data. *International Journal of Enterprise Information Systems*, *16*(4), 101–145. doi:10.4018/IJEIS.2020100106

Ustaoglu, E., & Aydinoglu, A. Ç. (2019). Theory, Data, and Methods. In H. Abusaada, C. Vellguth, & A. Elshater (Eds.), *Handbook of Research on Digital Research Methods and Architectural Tools in Urban Planning and Design* (pp. 156–202). IGI Global. doi:10.4018/978-1-5225-9238-9.ch009

Uttley, J., Simpson, J., & Qasem, H. (2018). Eye-Tracking in the Real World. In F. Aletta & J. Xiao (Eds.), *Handbook of Research on Perception-Driven Approaches to Urban Assessment and Design* (pp. 368–396). IGI Global. doi:10.4018/978-1-5225-3637-6.ch016

Vaitis, M., Kavroudakis, D., Koukourouvli, N., Simos, D., & Sarigiannis, G. (2019). An Integrated Information System for Road Traffic Accidents. *International Journal of Applied Geospatial Research*, *10*(3), 25–45. doi:10.4018/IJAGR.2019070102

Valadas, R. C., & Carvalho, E. S. (2020). Recommendation System for Sightseeing Tours. *International Journal of Tourism and Hospitality Management in the Digital Age*, *4*(2), 34–54. doi:10.4018/IJTHMDA.2020070104

Vargas-Hernandez, J. (2020). Regional Trade and Green Innovation Development Under the NAFTA. In A. C. Prabhakar, G. Kaur, & V. Erokhin (Eds.), *Regional Trade and Development Strategies in the Era of Globalization* (pp. 256–273). IGI Global. doi:10.4018/978-1-7998-1730-7.ch014

Vargas-Hernández, J. G. (2020). Rural Migration and Shrinkage Transformation Processes in Mexican Countryside. In D. Ushakov (Ed.), *Migration and Urbanization* (pp. 303–328). IGI Global. doi:10.4018/978-1-7998-0111-5.ch016

Vargas-Hernández, J. G. (2020). Strategic Transformational Transition of Green Economy, Green Growth, and Sustainable Development. In M. Merviö (Ed.), *Global Issues and Innovative Solutions in Healthcare, Culture, and the Environment* (pp. 99–119). IGI Global. doi:10.4018/978-1-7998-3576-9.ch006

Vargas-Hernández, J. G. (2020). Territory Democratization and Institutional Design in Urban Green Innovation. In Mapping, Managing, and Crafting Sustainable Business Strategies for the Circular Economy (pp. 95-110). IGI Global. doi:10.4018/978-1-5225-9885-5.ch006

Vargas-Hernández, J. G. (2020). Transition of Ecosystem Services Based on Urban Agro Ecology. In M. Merviö (Ed.), *Global Issues and Innovative Solutions in Healthcare, Culture, and the Environment* (pp. 14–33). IGI Global. doi:10.4018/978-1-7998-3576-9.ch002

Vargas-Hernández, J. G. (2020). Urban Governance, Democratic Decentralization, and Natural Resources. In S. Patti & G. Trizzino (Eds.), *Advanced Integrated Approaches to Environmental Economics and Policy* (pp. 175–199). IGI Global. doi:10.4018/978-1-5225-9562-5.ch009

Vargas-Hernández, J. G. (2020). Urban Sustainable Growth, Development, and Governance Structures for Revitalization of Open Vacant Spaces in Agriculture and Farming. In A. E. Obayelu & O. A. Obayelu (Eds.), *Developing Sustainable Food Systems, Policies, and Securities* (pp. 38–59). IGI Global. doi:10.4018/978-1-7998-2599-9.ch003

Vargas-Hernández, J. G. (2021). Green Resilience Eco-Oriented Land Uses in Urban Socio-Ecosystems. In A. A. Galaby & A. A. Abdrabo (Eds.), *Handbook of Research on Creative Cities and Advanced Models for Knowledge-Based Urban Development* (pp. 120–142). IGI Global. doi:10.4018/978-1-7998-4948-3.ch007

Vargas-Hernández, J. G. (2021). Sustainable Urban Agro Ecology and Its Implications With Food Systems. In Eco-Friendly Energy Processes and Technologies for Achieving Sustainable Development (pp. 143-166). IGI Global. doi:10.4018/978-1-7998-4915-5.ch008

Vargas-Hernández, J. G. (2021). Urban Socio-Ecosystems Green Resilience. In Eco-Friendly Energy Processes and Technologies for Achieving Sustainable Development (pp. 220-242). IGI Global. doi:10.4018/978-1-7998-4915-5.ch012

Vargas-Hernández, J. G., & Danish, M. S. (2021). System of Green Resilience Eco-Oriented Land Uses in Urban Socio-Ecosystems. In Eco-Friendly Energy Processes and Technologies for Achieving Sustainable Development (pp. 1-23). IGI Global. doi:10.4018/978-1-7998-4915-5.ch001

Vargas-Hernández, J. G., & Kyari, A. K. (2019). Economic and Environmental Effects of Rural and Urban Shrinkage Transformation Processes in Mexico. In U. G. Benna (Ed.), *Industrial and Urban Growth Policies at the Sub-National, National, and Global Levels* (pp. 191–208). IGI Global. doi:10.4018/978-1-5225-7625-9.ch010

Vargas-Hernández, J. G., & Pallagst, K. (2020). Green Innovation and Sustainable Urban Ecosystems. In B. R. Hernández-Sánchez, J. C. Sánchez-García, & A. C. Moreira (Eds.), *Building an Entrepreneurial and Sustainable Society* (pp. 54–74). IGI Global. doi:10.4018/978-1-7998-2704-7.ch003

Vargas-Hernández, J. G., Pallagst, K., & Zdunek-Wielgołaska, J. A. (2019). Urban Green Spaces for Sustainable Community Development. In U. G. Benna (Ed.), *Optimizing Regional Development Through Transformative Urbanization* (pp. 271–287). IGI Global. doi:10.4018/978-1-5225-5448-6.ch013

Vargas-Hernández, J. G., & Pallagst, K. M. (2020). Implications of Urban Sustainability, Socio-Ecosystems, and Ecosystem Services. In A. K. Rathoure & P. B. Chauhan (Eds.), *Current State and Future Impacts of Climate Change on Biodiversity* (pp. 31–53). IGI Global. doi:10.4018/978-1-7998-1226-5.ch003

Vargas-Hernández, J. G., & Zdunek-Wielgołaska, J. A. (2019). Entrepreneurship Development in the Urban Metropolitan Guadalajara. In O. L. Rua (Ed.), *Entrepreneurial Orientation and Opportunities for Global Economic Growth* (pp. 222–247). IGI Global. doi:10.4018/978-1-5225-6942-8.ch010

Vargas-Hernández, J. G., & Zdunek-Wielgołaska, J. A. (2019). Recent trends in Sustainable Urban Planning. *International Journal of Sustainable Economies Management*, *8*(3), 29–47. doi:10.4018/IJSEM.2019070103

Vasilikou, C. (2018). The Doors of Dynamic Thermal Perception. In F. Aletta & J. Xiao (Eds.), *Handbook of Research on Perception-Driven Approaches to Urban Assessment and Design* (pp. 182–205). IGI Global. doi:10.4018/978-1-5225-3637-6.ch008

Vaz de Freitas, I., Cortés, R. E., & Leite, P. P. (2019). A Methodological Proposal for Building Conservation. In S. Srivastava (Ed.), *Conservation and Promotion of Heritage Tourism* (pp. 57–86). IGI Global. doi:10.4018/978-1-5225-6283-2.ch003

Vaz do Vale, H. (2020). Profit and Profitability and a Look Through the Lens of Smart Cities. *International Journal of Entrepreneurship and Governance in Cognitive Cities*, *1*(1), 1–30. doi:10.4018/IJEGCC.2020010101

Verma, C., Pandey, R., & Katiyar, D. (2018). Performance Evaluating System Based on MapReduce in Context of Educational Big Data. *International Journal of Organizational and Collective Intelligence*, *8*(1), 1–12. doi:10.4018/IJOCI.2018010101

Vernikoff, L., Schram, T., Reagan, E. M., Horn, C., Goodwin, A. L., & Couse, L. J. (2019). Beyond Urban or Rural. In T. E. Hodges & A. C. Baum (Eds.), *Handbook of Research on Field-Based Teacher Education* (pp. 256–279). IGI Global. doi:10.4018/978-1-5225-6249-8.ch011

Vishkaie, R., & Levy, R. M. (2018). SketchBoard. In C. Nunes Silva (Ed.), *New Approaches, Methods, and Tools in Urban E-Planning* (pp. 223–250). IGI Global. doi:10.4018/978-1-5225-5999-3.ch008

Voghera, A., La Riccia, L., & Ardissono, L. (2019). Semantic Spatial Representation and Collaborative Mapping in Urban and Regional Planning. In A. Voghera & L. La Riccia (Eds.), *Spatial Planning in the Big Data Revolution* (pp. 194–220). IGI Global. doi:10.4018/978-1-5225-7927-4.ch009

Vujičić, T. M., & Simonović, D. (2020). Urban-Rural Synergy in Shrinking Regions Through Application of e-co Model. In A. Krstić-Furundžić & A. Djukić (Eds.), *Handbook of Research on Urban-Rural Synergy Development Through Housing, Landscape, and Tourism* (pp. 335–360). IGI Global. doi:10.4018/978-1-5225-9932-6.ch017

Vyas, S., & Bhargava, D. (2019). Big Data Utilization, Benefits, and Challenges for Smart City Implementation. In Handbook of Research on Big Data and the IoT (pp. 42-54). IGI Global. doi:10.4018/978-1-5225-7432-3.ch003

Waheed, A., & Shafi, J. (2019). Efficient Cyber Security Framework for Smart Cities. In R. A. Shaikh (Ed.), *Secure Cyber-Physical Systems for Smart Cities* (pp. 130–157). IGI Global. doi:10.4018/978-1-5225-7189-6.ch006

Walther, F., Weber, P., & Kabst, R. (2018). Analyzing Influencing Factors on Citizens' Trust and Satisfaction. *International Journal of Public Administration in the Digital Age*, *5*(2), 40–51. doi:10.4018/IJPADA.2018040104

Ward Mather, L., & Robinson, P. (2020). Durable Civic Technology. In C. N. Silva (Ed.), *Citizen-Responsive Urban E-Planning* (pp. 252–281). IGI Global. doi:10.4018/978-1-7998-4018-3.ch010

Wardell, T., Bevere, J., McCarty, J., Smith, W., Mulvaney, T., & Niesz, L. (2020). Transformational Leadership Initiatives Driving P-12 School Change. In J. R. O'Connor (Ed.), *Strategic Leadership in PK-12 Settings* (pp. 133–162). IGI Global. doi:10.4018/978-1-5225-9242-6.ch010

Weinstein, Z. (2020). How to Humanize Technology in Smart Cities. *International Journal of E-Planning Research*, *9*(3), 68–84. doi:10.4018/IJEPR.2020070104

Williams, Y. (2019). Building a Conceptual Framework for Culturally Inclusive Collaboration for Urban Practitioners. In B. M. Rice (Ed.), *Global Perspectives on Inclusive Teacher Education* (pp. 83–97). IGI Global. doi:10.4018/978-1-5225-7703-4.ch006

Wilson, B., & Cong, C. (2020). A Survey of Municipal Open Data Repositories in the U.S. *International Journal of E-Planning Research*, *9*(4), 1–22. doi:10.4018/IJEPR.2020100101

Wolf, K. L. (2020). The Urban Forest and Shopping Environments. In F. Musso & E. Druica (Eds.), *Handbook of Research on Retailing Techniques for Optimal Consumer Engagement and Experiences* (pp. 233–256). IGI Global. doi:10.4018/978-1-7998-1412-2.ch011

Wood-Harper, T. (2021). Future Teaching and Learning Applications in the Smart Campus. In M. B. Ali & T. Wood-Harper (Eds.), *Fostering Communication and Learning With Underutilized Technologies in Higher Education* (pp. 218–232). IGI Global. doi:10.4018/978-1-7998-4846-2.ch015

Woolliscroft, T. (2020). Defining Smart Community in the Context of Healthcare Efficiency in the UK. *International Journal of Intelligent Information Technologies, 16*(4), 1–18. doi:10.4018/IJIIT.2020100101

Xiao, J. (2018). Smell, Smellscape, and Place-Making. In F. Aletta & J. Xiao (Eds.), *Handbook of Research on Perception-Driven Approaches to Urban Assessment and Design* (pp. 240–258). IGI Global. doi:10.4018/978-1-5225-3637-6.ch010

Xu, T., Ma, Y., & Wang, Q. (2018). Cross-Urban Point-of-Interest Recommendation for Non-Natives. *International Journal of Web Services Research, 15*(3), 82–102. doi:10.4018/IJWSR.2018070105

Xue, F., & Gou, Z. (2018). Healing Space in High-Density Urban Contexts. In F. Aletta & J. Xiao (Eds.), *Handbook of Research on Perception-Driven Approaches to Urban Assessment and Design* (pp. 489–507). IGI Global. doi:10.4018/978-1-5225-3637-6.ch021

Yachba, K., Bendaoud, Z., & Bouamrane, K. (2018). Toward a Decision Support System for Regulation in an Urban Transport Network. *International Journal of Strategic Information Technology and Applications, 9*(2), 1–17. doi:10.4018/IJSITA.2018040101

Yadav, N., & Badal, N. (2020). Introduction to IoT Technologies and Its Applications. In R. Singh, A. Gehlot, V. Jain, & P. K. Malik (Eds.), *Handbook of Research on the Internet of Things Applications in Robotics and Automation* (pp. 238–264). IGI Global. doi:10.4018/978-1-5225-9574-8.ch012

Yadav, S., Reddy, B. E., & Srinivasa, K. (2018). A Centralized Health Monitoring System Using Health Related Sensors Integrated to the Mobile. *International Journal of Vehicular Telematics and Infotainment Systems, 2*(1), 68–79. doi:10.4018/IJVTIS.2018010105

Yavuz, N., Karkın, N., & Sevinç Çubuk, E. B. (2020). Explaining Government Crowdsourcing Decisions. In Digital Government and Achieving E-Public Participation (pp. 159-183). IGI Global. doi:10.4018/978-1-7998-1526-6.ch008

Yildirim, S., & Kaplan, M. (2020). Seafood Security and Sustainability Through Sustainable Development. In A. E. Obayelu & O. A. Obayelu (Eds.), *Developing Sustainable Food Systems, Policies, and Securities* (pp. 16–36). IGI Global. doi:10.4018/978-1-7998-2599-9.ch002

Yu, W., & Xu, C. (2018). Developing Smart Cities in China. *International Journal of Public Administration in the Digital Age*, *5*(3), 76–91. doi:10.4018/IJPADA.2018070106

Yuan, D., Yau, Y., & Bao, H. (2019). Institutional Innovation for Urban Village Renewal in Mainland China. In Q. R. Yasser (Ed.), *Handbook of Research on Rural Sociology and Community Mobilization for Sustainable Growth* (pp. 56–74). IGI Global. doi:10.4018/978-1-5225-7158-2.ch004

Yuan, S. (2020). *Business Reinvention Strategy*. IGI Global. doi:10.4018/978-1-7998-1550-1.ch008

Zawawi, A. A., Porter, N., & Ives, C. D. (2020). Greenways and Sustainable Urban Mobility Systems. In R. M. Doheim, A. A. Farag, & E. Kamel (Eds.), *Humanizing Cities Through Car-Free City Development and Transformation* (pp. 32–70). IGI Global. doi:10.4018/978-1-7998-3507-3.ch002

Zhang, W., Zou, W., & Qiu, X. (2019). A Unique Development Road of Urban Public Libraries of China. *International Journal of Library and Information Services*, *8*(2), 51–71. doi:10.4018/IJLIS.2019070104

Zhao, J., & Jin, Z. (2018). Predict Coordinated Development Degree of County Eco-Environment System Using GA-SVM. *Journal of Global Information Management*, *26*(3), 1–10. doi:10.4018/JGIM.2018070101

Zhou, C. (2020). Smart Creative Tourism. *International Journal of Urban Planning and Smart Cities*, *1*(1), 58–69. doi:10.4018/IJUPSC.2020010105

Zhou, L., Yao, Y., Guo, R., & Xu, W. (2021). Core Technology of Smart Cities. In K. Lyu, M. Hu, J. Du, & V. Sugumaran (Eds.), *AI-Based Services for Smart Cities and Urban Infrastructure* (pp. 47–90). IGI Global. doi:10.4018/978-1-7998-5024-3.ch003

Zhou, Y., & Xiong, Y. (2017). Live Broadcast Classroom. *International Journal of Distance Education Technologies*, *15*(3), 31–46. doi:10.4018/IJDET.2017070103

Zielonko-Jung, K., & Poćwierz, M. (2018). The Impact of Forms of Buildings on the Air Exchange in Their Environment. In E. Rynska, U. Kozminska, K. Zinowiec-Cieplik, J. Rucinska, & B. Szybinska-Matusiak (Eds.), *Design Solutions for nZEB Retrofit Buildings* (pp. 310–330). IGI Global. doi:10.4018/978-1-5225-4105-9.ch014

Zimmerman, A. S., & Herridge, A. S. (2019). Campus Climate and the Theory of Gender Performativity. In R. Jeffries (Ed.), *Diversity, Equity, and Inclusivity in Contemporary Higher Education* (pp. 226–242). IGI Global. doi:10.4018/978-1-5225-5724-1.ch014

Zschaber de Araújo, R. P., Moura, A. C., & Nogueira, T. D. (2018). Creating Collaborative Environments for the Development of Slum Upgrading and Illegal Settlement Regularization Plans in Brazil. *International Journal of E-Planning Research*, *7*(4), 25–43. doi:10.4018/IJEPR.2018100102

Zuma, K. P., Mason, R. B., & Maharaj, M. (2019). Awareness of the Space Industry as a Career Opportunity in South Africa. *International Journal of Sociotechnology and Knowledge Development*, *11*(1), 29–51. doi:10.4018/IJSKD.2019010103

About the Author

Patricia McKenna is the President of AmbientEase, providing information services to many sectors. Patricia works within and across diverse domains of scholarship and practice (interdisciplinary) and collaborates in team efforts to set up international, national, regional, and local information services, research projects, startups, and other creative and future-oriented initiatives. McKenna is also the Director of the UrbanitiesLab, an initiative focusing on the exploration of smart cities and learning cities. Author of some 50 publications that include contributions to special journal issues, books, and chapter contributions to books pertaining to the smart cities domain, Patricia is a regular presenter of papers at international conferences as well as a program board member and session organizer. With a focus on smart cities and learning cities, Patricia engages with diverse sectors around use experience and unexpected possibilities for generating and leveraging increased awareness and creativity in 21st century information environments. Patricia holds a BA from the University of New Brunswick, an MLS from McGill University, and a Doctorate in information management (with a focus on emerging technologies) from Syracuse University.

Index

A

C

D

E

G

I

L

N

O

P

S

T

U

V

W

www.ingramcontent.com/pod-product-compliance
Lightning Source LLC
LaVergne TN
LVHW081331300926
849161LV00008B/1939